P9-ELS-611

STEM CELL DIALOGUES

SHELDON KRIMSKY

STEM CELL DIALOGUES

A Philosophical
and Scientific Inquiry
Into Medical Frontiers

COLUMBIA UNIVERSITY PRESS
NEW YORK

Columbia University Press
Publishers Since 1893
New York Chichester, West Sussex
cup.columbia.edu

Library of Congress Cataloging-in-Publication Data
Krimsky, Sheldon.
Stem cell dialogues : a philosophical and scientific inquiry
into medical frontiers / Sheldon Krimsky.
pages cm
Includes bibliographical references and index.
ISBN 978-0-231-16748-2 (cloth : alk. paper) — ISBN 978-0-231-53940-1 (electronic)
1. Embryonic stem cells—Research—Moral and ethical aspects.
2. Genetics—Moral and ethical aspects. 3. Medical genetics—Moral and
ethical aspects. 4. Bioethics. I. Title.

QH588.S83K75 2015
174.2'8—dc23
2014039574

Columbia University Press books are printed on permanent
and durable acid-free paper.
This book is printed on paper with recycled content.
Printed in the United States of America

c 10 9 8 7 6 5 4 3 2 1

COVER DESIGN: Faceout Studio/Charles Brock
COVER IMAGE: © Getty

For my wife, Carolyn Boriss-Krimsky,
whose playwriting sparked my interest
in using dialogues in this book.

CONTENTS

ANNOTATED TABLE OF CONTENTS

ACKNOWLEDGMENTS

I wish to thank the following people who provided valuable advice and consultation at different stages of the book's development: Carlos Sonnenschein, Lee Silver, Stuart Newman, Shirley Shalev, Stuart Waldman, and especially Jonathan Garlick, who invited me to co-teach courses with him on science and social issues, where I was introduced to stem cells. I am indebted to Patrick Fitzgerald of Columbia University Press for his support and guidance throughout the development of this book, and to Leslie Kriesel for her superb copyediting.

INTRODUCTION

The term "stem cell," first introduced in the backwaters of cell and developmental biology nearly 150 years ago, has recently become a central organizing concept in a rapidly growing field of biomedical science. And because of their association with the human embryo, stem cells have also entered the lexicon of popular culture in a maelstrom of controversy. They have come to signify a new frontier called "regenerative medicine." Stem cells are understood as the master building blocks and regenerating cells of an organism, starting in the fertilized egg and continuing in some form throughout development. Scientists have predicted that once stem cells can be isolated, cultured, and decoded so they can be differentiated into specialized cells, damaged brain, nerve, and heart cells can be repaired just as the body repairs itself when you cut your finger. Through a series of dramatic dialogues involving fictitious characters, this book explores the scientific, social, and ethical dimensions of stem cell science—one of the most dynamic and promising research programs in modern biology.

The history of biomedical science is punctuated by Nobel Prize-worthy discoveries that have had transformative, multidisciplinary effects on scientific research communities. In 1953 the major discovery was the structure of the DNA molecule by James Watson and Francis Crick. Twenty years later it was Stanley Cohen and Herbert Boyer's pathbreaking work on a new technique called recombinant DNA molecule technology, by which they were able to transfer functioning genes from one organism to another, even across species.[1] In 1975 Frederick Sanger and Alan Coulson discovered a rapid method for determining sequences in DNA.[2] This led to the sequencing of the human genome twenty-five years later and created both a scientific and a commercial tidal wave of activity in genomics, bringing thousands of scientists into this work. By 1983, Kary Mullis discovered the polymerase chain reaction (PCR), an innovative new chemical technique used to rapidly copy a single strand (or a few strands) of DNA, generating millions of copies of the DNA sequence. This technique has been applied to many areas of medicine, such as the diagnosis of infectious and hereditary diseases, and to physical anthropology. It has also revolutionized forensic DNA profiling in criminal investigation, which began in England in the 1980s with the use of restriction fragment length polymorphism, an earlier and more cumbersome method of using DNA sequences for identification.[3]

A more recent transformative event in modern biology took place at the University of Wisconsin in 1998. James Thomson isolated and cultured the first human embryonic stem cells from donated early stage human embryos. His discovery was built on the achievements of predecessors including Sir Martin J. Evans, who was awarded the Nobel Prize in Physiology or Medicine for isolating embryonic stem cells in mouse blastocysts and using them to produce transgenic mice (a transgenic animal has had genes from another species transferred to its genome). Thomson's discovery created an international tsunami in cell biology. The new operative phrase was "regenerative medicine." When certain cells in the body are damaged or die, they do not regenerate themselves to create new tissue. This is true of brain cells (after a stroke), nerve cells (after spinal cord injury), heart cells (after a heart attack), and retinal cells (from macular degeneration). Human embryonic stem cells are the precursors for all cells in the body. Therefore, if those cells could be harvested and given the right signals, then the differentiated cells they produce, when introduced into the right location in the body, could in theory regenerate new cells and repair damaged tissue. The excitement that ran through the scientific and biomedical community was palpable. New centers of regenerative medicine were established. Voters in the state of California passed an unprecedented bond issue to fund a stem cell institute that circumvented the federal legal quagmire over using human embryos for developing cell lines. The fields of cell and tissue biology were reinvented as the new frontier in biomedical research for the twenty-first century. And while genetics certainly played a role, the proof of concept was in getting stem cells to do, under scientific direction, what they do naturally in the normal development of an organism.

The stem cell research program faced many social obstacles—not the least of which was the new demand for embryos from which to harvest the embryonic stem cells. The scientific and public responses became inextricably intertwined. The social issues faced by stem cell scientists were different than those of the early recombinant DNA scientists. Geneticists in the 1970s dealt with concerns about whether gene-splicing experiments might introduce new pathogens. And although a small minority of voices challenged the use of gene splicing as unnatural, such concerns had an imperceptible impact on the progress of the field.

In contrast, stem cell science met society head on almost immediately over the issue of destroying human embryos to obtain the highly valued stem cells. The public and scientific discourses took many different trajectories. The complexity of the wide-ranging discussions included the hopes and dreams of individuals who were given new reason to believe they could be cured, like Christopher Reeve, the actor superhero who became paralyzed after he fell from a horse. He considered stem cell therapy the best hope for treating his spinal cord injury. The popular cinema and TV actor Michael J. Fox embraced the new research in stem cells as a promising treatment strategy for his own Parkinson's disease.

When the administration of George W. Bush put limits on federal funding for human embryonic stem cell research, the policy acted like an accelerant, feeding the passions of those who believed that biomedical science had reached a new plateau that would forever change the treatment of chronic, incapacitating, and life-threatening diseases. The obstacles set by Congress and the president made stem cell advocates more aggressive in raising private money and winning public acceptance.

In co-teaching several courses on "stem cells and human enhancement" with two cell biologists engaged in stem cell research and a philosopher, I observed a rich tapestry of social and scientific discourses that engaged ethics, science, religion, social media, government, law, and economics. I thought about how best to present these issues to a general audience in an accessible but scientifically sound way, and was reminded of the dialogues I read episodically throughout my education. One of the oldest applications of the dialogue form was by the Greek philosopher Plato, who used his teacher Socrates as the main character. The "Socratic Dialogues" illustrate the conflict between Plato's philosophy and the values that were prevalent in his society during the fifth century B.C. in Athens. Plato's incorporation of the dialogue form was an affirmation that philosophical inquiry is a social enterprise. When people are confronted with alternative viewpoints, they have an opportunity to sharpen, correct, or retract their original position. His dialogues also invite readers to take sides or to re-examine their own beliefs: "the individual's own point of view on a problem only emerges as it comes into conflict with the points of view expressed by other individuals participating in the dialogue."[4]

Plato uses Socrates as the principal protagonist to cross-examine his main adversaries, to shake their confidence in their claims of authority on issues of piety, justice, or knowledge itself. Ultimately, his dialogues demonstrate that his adversaries do not know what they pretend to know. And his main protagonist, Socrates, expresses his obligation to question Athenians and affirms that "the unexamined life is not worth living." Here is a short segment from Plato's *The Meno*, where the personages Meno and Socrates discuss the difference between opinion and knowledge.[5]

MENO: What do you mean by the word "right"? [as in being correct]

SOCRATES: I will explain. If a man knew the way to Larisa, or anywhere else, and went to the place and led others thither, would he not be right and a good guide?

MENO: Certainly.

SOCRATES: And a person who had a right opinion about the way, but had never been and did not know, might be a good guide also, might he not?

MENO: Certainly.

SOCRATES: And while he has true opinion about that which the other knows he will be just as good a guide if he thinks the truth, as he who knows the truth.

MENO: Exactly.

SOCRATES: Then true opinion is as good a guide to correct action as knowledge; and that was the point which we omitted in our speculation about the nature of virtue, when we said that knowledge only is the guide of right action, whereas there is also right opinion.

MENO: True.

SOCRATES: Then right opinion is not less useful than knowledge?

MENO: The difference, Socrates, is only that he who has knowledge will always be right; but he who has right opinion will sometimes be right, and sometimes not.

. . .

SOCRATES: While [true opinions] abide with us they are beautiful and fruitful, but they run away out of the human soul, and do not remain long, and therefore they are not of much value until they are fastened by the tie of the cause. . . . But when they are bound, in the first place, they have the nature of knowledge; and, in the second place, they are abiding. And this is why knowledge is more honorable and excellent than true opinion, because fastened by a chain.[6]

As Plato conceived it, the dialogue is a literary form that gives the philosopher complete control over his argument. At times he allows Meno to ask questions and offer conclusions, but these are always consistent and in agreement with Plato's views as expressed through the voice of Socrates. The reader is never in doubt about who is in charge.

Another historically significant dialogue is Galileo's *Dialogues Concerning Two New Sciences.*[7] Galileo selected advocates for two opposing views who speak in their local idiom and present contrasting scientific theories that distinguished his own views from those of his Aristotelian opponents.[8] Simplicio represents the Aristotelian and Ptolemaic theory of the universe. Salviati presents the modern view via the Copernican theory, which Galileo supported. A third character in the *Dialogues* is Sagredo, representing an educated member of the public, cast as an unbiased observer responding to the adversaries.

Galileo used the dialogue form as a means to communicate his criticism of Aristotle's generally accepted theory of moving bodies, that heavier bodies move faster than lighter bodies. In the following extract, Galileo frames a clever thought experiment in the voice of Salviati to refute Aristotle in the form of a logical contradiction. Galileo is the bridge between ancient and modern science. He shows us through the *Dialogue* that empirical observations are not the only means to refute a scientific claim and that "thought experiments" can play an important role in science.[9]

SALVIATI: But, even without further experiment, it is possible to prove clearly, by means of a short and conclusive argument, that a heavier body does not move more rapidly than a lighter one, provided both bodies are of the same material and in short such as those mentioned by Aristotle.

But tell me, Simplicio, whether you admit that each falling body acquires a definite speed fixed by nature, a velocity which cannot be increased or decreased except by the use of force or resistance.

SIMPLICIO: There can be no doubt that one and the same body moving in a single medium has a fixed velocity determined by nature and which cannot be increased except by the addition of momentum or diminished except by some resistance that retards it.

SALVIATI: If we then take two bodies whose natural speeds are different, it is clear that on uniting the two, the more rapid one will be retarded by the slower, and the slower will be somewhat hastened by the swifter. Do you not agree with me in this opinion?

SIMPLICIO: You are unquestionably right.

SALVIATI: But if this is true and if a large stone moves with a speed of, say, eight, while the smaller stone moves with the speed of four, then when they are united, the system will move with a speed of less than eight; but the two stones when tied together make a stone larger than that which moves with the speed of eight. Hence, the heavier body moves with less speed than the lighter; an effect which is contrary to your supposition. Thus, you see how from your assumption that the heavier body moves more rapidly than the lighter one, I infer that the heavier body moves more slowly.

Galileo's personages, like those in Plato's dialogues, are not equal intellectual rivals. One is usually a "straight man" who, while questioning Galileo's views, is quite compliant with his responses.

My third example is the French philosopher Nicolas Malebranche, who in 1688 composed *Dialogues on Metaphysics and on Religion* and revived the literary genre for philosophy. In this extract from Malebranche's *Philosophical Selections*,[10] Aristes and Theodore discuss reality and illusion as Malebranche builds his metaphysical foundation for knowledge.

THEODORE: Ah, my dear Aristes, once again watch out that I do not go astray. . . . Also, you will learn nothing if it is not your own reflections which put you in possession of the truths that I shall attempt to demonstrate. There are only three sorts of Beings of which we have any knowledge and to which we can have some connection. God, or infinitely perfect Being, is the principle or cause of all things; minds, which are known only through internal consciousness (sentiment), which we have of our nature; and bodies, of the existence of which we are assured by the revelation we have of them. Now, what we call a man is simply a composite.

ARISTES: Not so fast, Theodore. I know there is a God or infinitely perfect Being. For, if I think of such a being, which I certainly do, that being must exist, since nothing finite can represent the infinite. I also know that minds exist on the assumption that there are beings resembling me. For I cannot doubt that I think, and I know that what thinks is something other than extension or matter. You have proved these truths to me. But what do you mean by our being assured of the existence of bodies "by the revelation we have of them"? What! Do we not see them? Do we not feel them? We do not need "revelation" to teach us that we have a body when we are pricked: we very truly sense it.

THEODORE: Yes, no doubt we sense it. But the sensation of pain we have is a kind of "revelation." . . . You still forget that it is God Himself who produces in our souls all the different sensations affecting them on the occasion of change happening to our bodies and in consequence of general laws . . . which are simply the uniform efficacious volitions of the Creator. . . . The point which pricks the hand does not pour pain into the hole which it makes in the body. . . . It is then God Himself who, by sensations with which He affects us, reveals to us what is happening outside us.

ARISTES: I was wrong, Theodore. But what you are saying brings to mind a very strange thought. . . . It is that I am beginning to doubt the existence of bodies. The reason is that the revelation which God gives us of their existence is not sure. It is after all certain that we sometimes see what does not exist, for instance, when we are asleep or when fever causes a disturbance in our brains. If God can, as you say, sometimes give us deceptive sensations in consequence of His general laws, if through our senses, He can reveal to us things that are false, why can He not always do this, and how then will we be able to distinguish truth from falsity in the obscure and confused testimony of our senses? It seems to me that prudence dictates that I suspend judgment with regard to the existence of bodies. Please give me an exact demonstration of their existence.

THEODORE: "An exact demonstration"! That is a bit much, Aristes. I admit I do not have one.

As illustrated in Malebranche's dialogue, philosophical discourse about the fallibility of the senses and the foundations of knowledge became a central focus of epistemology and metaphysics from the Enlightenment to modern times. These three examples of dialogues illustrate a form of advocacy on the part of the authors.

In *Stem Cell Dialogues*, I chose to create more balanced discussions and interrogations among people who are on a relatively equal intellectual footing. The central character is Rebecca Franklin, an M.D., Ph.D medical geneticist with a background in bioethics. Dr. Franklin has a quadriplegic father and has redesigned her career from medical genetics to stem cell research in order to develop a therapy to cure her father's condition.

The dialogue literary form in the way I have constructed it has some unique values for communicating scientific, social, and ethical ideas. It permits the writer to create tension in an argument between two protagonists and to raise "simple" questions and responses that introduce readers to a new scientific discovery. The dialogue form can allow an argument to develop while exploring epistemological and ethical issues through different voices. By creating interactions between characters, I am able to highlight the human and social dimensions of science. *Stem Cell Dialogues* consists of twenty-five dialogues, each with fictitious personages discussing one of the many themes of stem cell science, including its social effects and ethical dimensions. The basic narrative in each dialogue is grounded on scientific studies and media reports about the issues discussed. The notes provide partial access to that record. While these *Dialogues* refer to real historical events, I chose to explore the events, and the people and companies who shaped them, through the voices of fictitious characters.

None of these characters depicts an actual person, and any similarities between the fictional characters and actual people are purely coincidental. The characters are merely a literary device used to explore the challenging scientific and social issues posed by stem cell research, and their statements and positions should not be construed as the actual statements or positions of any individual corporation or other entity. Further, I use these fictional characters for pedagogical purposes only and do not purport to speak on behalf of the corporations or organizations they are depicted as representing.

HARNESSING STEM CELLS FOR REGENERATIVE MEDICINE

Every species, from bacteria to mammals, is capable of some regeneration. Many invertebrates, including complex ones such as starfish, some insects, crabs, crawfish, and lobsters, are able to regenerate lost limbs. Some amphibians and reptiles can replace a lost leg or tail. Salamanders can regenerate limbs, tails, jaws, eyes, and some internal structures. Fish can regenerate parts of the brain, eye, kidney, heart, and fins. Mammals have a limited capacity to repair their bodies when they have been afflicted with disease, exposed to physical injury, or simply deteriorated from age, and humans have a small capacity compared to other species. We can regenerate some skin, a sizable amount of liver, blood, and the tips of fingers and toes. Severed digits or limbs do not regenerate. Neurons in the spinal cord damaged by injury, brain cells destroyed by lack of oxygen, and heart cells and other tissue damaged by lack of blood are not replenished by the body's repair mechanisms.

Biomedical and prosthetic engineers have invented replacements for all sorts of body parts that have failed, been severely damaged, or been amputated. Remarkable technological advances in artificial limb development have progressed so far that those who use them are thought to hold an advantage over able-bodied athletes in some sporting competitions. Artificial skin products for burns and nonhealing wounds are grown in a laboratory. When burns exceed the body's natural ability to repair itself, artificial skin replacements protect victims from deadly infections and activate alternative paths of skin regeneration.

Dialysis machines allow people who have lost kidney function to live and work. Cochlear implants give people with severe hearing loss the opportunity to hear again. Metallic rods and titanium hips affordepeople with bone degeneration normal lives. Bone marrow transfusions can cure patients with forms of leukemia.

Because the human body has not evolved the capacity to repair certain damaged cells, tissue, limbs, or organs and no complex functions that have become inactivated can be restored through DNA, scientists have pursued another path. They predict that by isolating, culturing, and differentiating embryonic stem cells, they can make any of the more than 200 kinds of somatic cells in the human body. If adult stem cells could be derived from the body's own somatic cells or from embryonic stem cells, we would have a truly regenerative body that could be repaired by transplanting new cells developed in the laboratory.

The discovery in 1998 that human embryonic stem cells can be used in this way gave scientists hope that a single cell line can be called into service to regenerate any damaged tissue or cell in anyone's body. Such power in a precursor cell could surpass the greatest medical discovery to date, broad spectrum antibiotics, which has saved countless lives from virulent bacteria.

The term "stem cell" is reported to have appeared in the scientific literature as early as 1868, in the writings of the German evolutionary biologist Ernst Haeckel. He used the term in one sense to denote the single-celled ancestor of the multicellular organism. "The name 'stem cell' seems to me the most simple and appropriate one, because all other cells stem from it and because it is in its most literal sense the stem father as well as the stem mother of all the countless generations of cells of which later on the multicellular organism is composed."[1]

At the turn of the twentieth century other scientists, most prominently Alexander Maximow, Vera Danschakoff, Ernst Newman, and Artur Pappenheim, wrote about stem cells in the blood system.[2] Maximow, a distinguished Russian histologist and embryologist, studied at St. Petersburg, Freiburg, and Berlin, then became a professor in St. Petersburg. After the Russian revolution he emigrated to the United States and accepted a professorship of anatomy at the University of Chicago. In 1908 Maximow used the term "stem cell" (*Stammzelle*) in a scientific paper he delivered at the Congress of the Hematologic Society in Berlin, where he postulated that all blood cells develop from a common precursor cell. From this idea he formulated the unitarian theory of hematopoiesis.[3]

Throughout the early part of the twentieth century, scientists intensified their study of blood cells to determine which had the regenerative capacity that had been postulated by Maximow and others. Studies of bone marrow in animals yielded cells that could regenerate and differentiate into several types of blood cells. A group of researchers at the Rockefeller Institute for Medical Research reported in 1936 their discovery of stem cells in the bone marrow of rabbits. The authors wrote: "The second question at issue in hematology concerns the nature of the stem cell. It is accepted that there is a common stem cell for all the white blood cells . . . this so called primitive cell occurs in bone marrow diffusely scattered and not in germ centers."[4] By the mid-1950s hematopoietic (blood) stem cell transplantation had begun. The multipotent (capable of producing a lineage of cells) hematopoietic stem cells are found in bone marrow and used to treat blood diseases like thalassemia, leukemia, sickle cell anemia, and aplastic anemia.

George Mathé, a French oncologist, performed the first European bone marrow transplant on five Yugoslavian nuclear plant workers whose bone marrow had been damaged by an accident. Although these first transplants did not work, Mathé succeeded in using transplants to treat leukemia patients. And in the United States, E. Donnall Thomas, at the Fred Hutchinson Cancer Research Center, received the Nobel Prize in Physiology or Medicine for his work in establishing bone marrow transplantation for the treatment of leukemia in the late 1950s. In a fifty-year retrospective of his contributions in the *New England Journal of Medicine*, Frederick Applebaum noted, "Thomas elucidated the nature of hematopoiesis and stimulated efforts to identify the hematopoietic stem cell and factors that control its growth and development."[5] The first person to perform a successful human bone marrow transplant to treat a disease other than cancer was Robert A. Good of the University of Minnesota.

STEM CELLS AND BONE MARROW TRANSPLANTS

Historical accounts of investigations into stem cells attribute their starting point to the aftermath of the atomic bombing of Hiroshima and Nagasaki. Many of the survivors had radiation damage that affected their ability to make new blood cells. Their bodies could not make new white blood cells for protection against infections or produce enough blood-clotting platelets. Scientists began studying the effects of radiation on mice and found that the mice died from blood diseases within weeks of exposure to whole-body radiation. By 1956, a team of scientists lead by Charles Ford, who was working at the Atomic Energy Research Establishment near Oxford, England, learned that when irradiated mice were injected with bone marrow from unexposed mice, they regenerated blood cells. By inference, scientists determined that something in the bone marrow had the capacity to regenerate healthy blood cells in an organism whose cells had been damaged by radiation. There were debates over whether it was a chemical in the marrow or a progenitor stemlike cell. Ford showed that it was certain stem cells that were capable of regenerating blood cells.[6]

Robert Alan Good was born in Crosby, Minnesota, in 1922. He completed his B.A. degree at the University of Minnesota and was the first student to complete the M.D.-Ph.D. joint degree at the university in 1947, a mere three years later. After a fellowship year at the Rockefeller Institute for Medical Research, he returned to the University of Minnesota to do research on the immune system. In 1968 he led a team that performed the first successful bone marrow transplant on a five-month-old boy with a severe immune deficiency.[7] Earlier attempts, in 1957 by E. Donnall Thomas, had not succeeded. Years later it became more fully understood that bone marrow contained stem cells that, when transplanted into a person whose blood cells were genetically or environmentally damaged, would produce a new line of blood cells.

By the third quarter of the twentieth century the most extensively studied stem cells were the hematopoieic cells (HPCs) present in blood and bone marrow. HPCs are capable of forming red blood cells, platelets, and white blood cells that fight infections. Some scientists even believed that the bone marrow stem cells were pluripotent, capable of generating cells outside of the blood system. Catherine Verfaille at the University of Minnesota grew bone marrow cells in culture and mixed them with cells like neurons on the hypothesis that the hematopoietic stem cells would be signaled by the neurons to make more of them. Verfaille reported that she had evidence that her hematopoietic stem cells in different cell cultures began to behave like heart, brain, and liver cells. There was a lot of skepticism when scientists could not replicate her results. A panel of experts commissioned by the University of Minnesota found that the process Verfaille had used to identify tissues from adult stem cells was flawed, which made her interpretation incorrect. Currently, neither blood stem cells nor any other type of adult stem cells has been found to be pluripotent.

Despite the fact that stem cells were sought, investigated, and used therapeutically throughout the twentieth century, it wasn't until the turn of the new millennium that the ethics of their use exploded as an issue in the United States. This followed the major breakthrough that took place in 1998 at the laboratory of James Thomson of the University of Wisconsin.

Thomson was born in 1958 in Oak Park, Illinois. He earned his B.S. degree in physics at the University of Illinois in 1981 and a doctorate in veterinary medicine and a doctorate in molecular biology at the University of Pennsylvania in 1985 and 1988 respectively. Thomson understood that mouse stem cells and human stem cells behaved differently and required different culturing techniques. He turned his attention to primates. In 1995 he performed the first isolation of embryonic stem cell lines from a nonhuman primate. And on November 6, 1998, the journal *Science* published a paper from Thomson's group at the Wisconsin Regional Primate Research Center at the University of Wisconsin reporting that they had isolated and cultured human stem cells. The cells came from donated excess embryos from couples who had contracted with an in vitro fertilization (IVF) clinic. Thomson's cells had survived, multiplied, and differentiated into the cells of the major human tissues for eight months. A scientific revolution in regenerative medicine was born.

Ole Johan Borge describes the sea change: "The announcement of the construction of pluripotent stem cell lines from human blastocysts [Thomson et al. 1998] and aborted fetuses [Shamblott et al. 1998] placed stem cells at the center stage of medical research and marked the starting point of a large and emotional bioethical debate."[8] Pluripotent stem cells were immediately seen as modern medicine's Holy Grail, holding the promise to treat every disease caused by lack of any given cell type.[9] With the isolation and culturing of embryonic stem cells, derived from IVF discarded embryos, the link between embryo and stem cell became forever established. And with that, the ethical controversy began.

The series of dialogues in this volume offers a glimpse of the many intersecting conversations taking place within and across the scientific, religious, bioethics, business, and political communities that debated or sought moral enlightenment about the use of early (three- to six-day-old) embryos to create stem cells that were considered destined to transform regenerative medicine.

Stem cell conversations dominated scientific communications, blogs, media networks, investment houses, universities, places of worship, and discussions among stakeholder groups of all stripes. Within the conversations could be found a broad spectrum of opinions and arguments. For some, human embryos were "sacred life forms" endowed by their creator. For others, early embryos were a bundle of undifferentiated cells. Even with these stark contrasts, there is above all a search for a higher, more enduring truth.

PRINCIPAL CHARACTER

In this book, the main character in this pursuit of scientific and moral truth is Dr. Rebecca Franklin, whose intellectual and personal stake in this journey is the driving force behind the narratives. She is a forty-two-year-old, auburn-haired, no nonsense medical geneticist whose achievements she would attribute to her ultrasupportive family and her precocious inquisitiveness, recognized by her parents when she was barely age two. She grew up in a lower middle-class household in Brooklyn, as an only child showered with her parents' affection. She turned their forsaken dreams of higher education for themselves into her own obsession.

Rebecca attended New York's prestigious Stuyvesant High School, long viewed as one of the nation's premier science high schools for men. It had become co-ed seventeen years before her enrollment, which helped to increase its national rankings. She devoured courses in biology, chemistry, and physics, mastered calculus and differential equations, and graduated at age sixteen.

Her father, Samuel, was a carpenter and small-time contractor who worked primarily on two-family homes that dotted the neighborhoods of Brooklyn and Queens. His wife, Denise, was a preschool teacher whose modest earnings supplemented her husband's income.

After Rebecca graduated Stuyvesant, she received a scholarship to Brown University, where she constructed a pre-med curriculum that consisted of heavy doses of genetics and bioethics. By the time she entered Brown, the new genetic technology had been active for twenty years, enough time to mature. Franklin developed skills as a bench scientist, acquiring basic techniques of recombinant DNA molecule technology and laboratory methods to study mutations in medically significant DNA sequences. By the time she graduated, she had set her sights on an M.D.-Ph.D. degree and a research career in medical genetics, where she hoped to study the genetics of autism. She was admitted to Harvard Medical School on a partial scholarship and an internship in medical ethics, and attended seminars on death and dying, patient autonomy, and justice in health care. After completing her medical degree, she spent another two years in one of Harvard's best medical genetics labs doing full genome studies, comparing the genes of autistic and nonautistic children and searching for candidate genes associated with autism. Dr. Franklin had one more academic road to travel. She enrolled in a master's degree program in bioethics at the University of Pennsylvania under the mentorship of America's foremost bioethicist, Arthur Caplan.

During her studies at Penn, Dr. Franklin's father had a near fatal accident, falling off a thirty-foot ladder while inspecting a roof on a contracting job. He damaged his spinal cord and became a quadriplegic with movement only in two fingers of his left hand. She was devastated by the tragedy. Her father had seemed invulnerable—tall, muscular, and able to repair almost anything, from automobile engines to home heating systems.

Not one to accept fatalism or destiny, Dr. Franklin believed that science could solve any medical problem if the resources were sufficient and properly focused. She saw the movie *Conviction*, starring Hilary Swank, in which the trailer-park sister of a convicted killer turns her life around to complete her GED and then enters law school in order to give her brother proper representation to prove his innocence. Franklin imagined herself entering a new field of medical research that was relevant to finding a cure for her father's paralysis. So she rearranged her career path and began working alongside one of the leading stem cell scientists in the country during a period when there was a groundswell of interest in the use of embryonic stem cells for regenerative medicine.

Dr. Franklin wanted not only to understand and advance the science of stem cells but also to embrace the ethical and social dimensions of using embryos to cure diseases. She came face to face with ideologues of various stripes, passionate researchers, diffident politicians, medical entrepreneurs, and individuals with life-threatening illnesses grasping at a thread of hope.

When she was in her laboratory, her mind fixated on the microscopic cells that filled her slides and culture dishes, the cells removed from three- to four-day-old embryos. Her job was to induce the embryonic cells to become specialized cells called oligodendrocytes that could eventually be delivered to the spinal cord of a paralyzed individual like her father. But the lab bench was involved in a public controversy. So she embarked on another journey that took her through a series of conversations and dialogues. It began with her conversation with her father.

STEM CELL DIALOGUES

DIALOGUE 1

HOPE

Scene: Sitting in a park on an April afternoon, Samuel Franklin, a sixty-six-year-old retired contractor, and his daughter, Rebecca, a physician and Ph.D. medical geneticist, are enjoying a conversation. Samuel is in a wheelchair and tethered to a breathing machine. Classified as a quadriplegic, he has partial movement in the fingers of one hand, enough to move a computer mouse. Rebecca is sitting next to him, on a bench.

REBECCA: What a glorious day. The sun feels like a radiant message on my back. [*After pausing, and in a somber tone*] I know today is the fifth anniversary. How do you feel?

SAMUEL: [*Propping his head up slightly with his finger motion on a computer mouse and speaking in a staccato voice, while taking deep breaths of oxygen*] That day is on constant replay. The moment never leaves me. The sun was magical then too. My life would be so different if only the ladder was properly secured! I dream of redoing the day a hundred million times, but when I wake up, nothing has changed.

REBECCA: But something *has* changed. Scientists have discovered a possible way of repairing damaged cells, like your nerve cells. They are learning to unleash the power of stem cells.

SAMUEL: I've heard something about stem cells. What are they?

REBECCA: Scientists have isolated cells found in early human embryos that have unique properties. They can be used to regenerate any cells or tissues in the body that have been damaged or destroyed. They're called embryonic stem cells. Your paralysis was caused by damaged cells in your spinal cord. Stem cells can also be personalized, so when they're transplanted into individuals they won't be rejected. These stem cells or their equivalents are the best new hope for helping people like you with spinal cord injury.

SAMUEL: My neurologist told me that my fall permanently damaged my nerve cells and they can't be replaced. They call it thoracic spinal cord injury.[1] All the physical therapies have not helped. To be honest, Rebecca, I've been through excruciating physical therapy, and nothing has helped reduce my paralysis. I have nothing left to live for. Without hope, I have no spirit left.

REBECCA: That's the point. These embryonic stem cells can be made to generate other precursor cells that leading neuroscientists believe can restore your damaged nerve cells. There is a great deal of research beginning in this field, and a lot of intelligent scientists are working on it.

SAMUEL: I've heard that there are all sorts of restrictions on using embryonic stem cells and that the government will only permit certain cell lines. Even "Superman," Christopher Reeve, couldn't change those policies for his own spinal cord injury.

REBECCA: There have been restrictions on the use of embryonic stem cells. But even if those remain, they only apply under government funding. There are many private sources funding the research, as well as other national governments. And one more thing: there is a new technology called induced pluripotent stem cells, or iPS cells. It changes your adult stem cells back into more primitive and powerful cells that can be used to make replacements for your damaged cells. Scientists have used iPS-derived cells to treat blood and neurological disorders in rats and mice. You have to have faith.

SAMUEL: I have faith in you, Rebecca. But it is difficult for me to have faith that my body can be repaired so that I can walk again or move my arms. If I'll never be able to do much more than move my second finger, I want you to help me out of this torment.

REBECCA: There have already been some promising animal experiments. Embryonic stem cell-derived neurons were injected into mice that had Parkinson's disease, and those cells turned into dopamine-producing neurons that reversed the symptoms of Parkinson's.[2] Also, cells derived from human embryonic stem cells were injected into paralyzed rats,[3] and the rats were able to walk again.[4] This highly publicized experiment even appeared on YouTube, and it's turned a lot of skeptics into true believers.[5] It won't be long before one of these therapies will be used on humans.[6]

SAMUEL: But it could take decades before they begin experimenting on humans.

REBECCA: Not so. The FDA has already given approval for the first clinical trial, where stem cells will be used to treat spinal cord injury. It's sponsored by the Geron Corporation. They are testing it on patients who had a complete thoracic spinal cord injury like yours.[7] And there will be many more trials. The science is in its infancy, and new discoveries are being made every day. We have a lot to be optimistic about.

SAMUEL: I don't have time to wait. Each day that I'm in this condition, I'm closer to getting someone to remove the breathing tube. It's my right to decide when the quality of my life is good enough to justify remaining alive and when it's not. Living is not the ultimate value when you face unbearable suffering.

REBECCA: Dad, you have given up so much to make it possible for me to be who I am. Give me an opportunity to give you a reason to live! We're beginning to do this work in my lab. I am redesigning my research career from searching for genetic determinants for disease to the problem of regenerative medicine, to produce cells like motor neurons that can cure your spinal cord injury. No one says finding a cure for paralysis will be easy. But there are thousands of people from many countries working on this approach. Getting a man on the moon wasn't easy either. But the collective efforts of dedicated and creative scientists were able to beat the odds. Hope is what sustains and propels scientific inquiry toward success. Progress is being made every day. I am confident there will be a breakthrough.[8]

SAMUEL: Hope can only sustain me if my minute-to-minute existence has more value than the alternative. At the moment that's true, but only barely, and time will change that—at which point I'll need your help to exit.

DIALOGUE 2

WHY IS THIS CELL DIFFERENT FROM OTHER CELLS?

The idea behind stem cells goes back to the 1800s, when scientists hypothesized that some cells were precursors to other, more differentiated cells. By the early 1900s, the first blood stem cells were postulated by the Russian biologist Maximow and others. A major breakthrough came during the late 1950s and early 1970s, when James Till and Ernest McCulloch collaborated at the Ontario Cancer Institute in Toronto in studying the effects of radiation on the bone marrow of mice. In 2005 they were awarded the prestigious Albert Lasker Basic Medical Research Award, which honors scientists whose contributions to research are of unique magnitude and have immeasurable influence on the course of science, health, or medicine, "for ingenious experiments that first identified a stem cell—the blood-forming stem cell—which set the stage for all current research on adult and embryonic stem cells."[1] The award description noted: "By the early 1970s, Till and McCulloch's experimental observations were clear-cut: They revealed that bone marrow transplantation owes its restorative powers to a single type of cell that not only can divide, but can differentiate into all three types of mature blood cells—red cells, white cells, and platelets. These features meant that the colony-forming cells represented a new class of progenitor cells—ones that could proliferate enough to repopulate the bone marrow of an entire animal, self-renew, and give rise to specialized cells that have limited life spans."[2]

In the late 1960s, scientists performed the first successful human bone marrow transplants to replenish blood cells. The stem cells in the transplanted marrow produced healthy blood cells in patients who had abnormal blood cells. This was an autologous stem cell transplant, in which the person from whom the cells were taken and the person treated were the same individual. Several years later the first bone marrow transplant was performed on unrelated people.

Animal embryos develop from fertilized eggs called zygotes. Up to four days of development, a zygote develops into a cluster of cells that are unique, and the most powerful in the life of the organism. They are called totipotent cells. These cells have the capacity to self-renew and to generate specialty cells, those that compose different tissues and organs. Not only do the totipotent cells, which are clustered in a solid mass in the early embryo called the morula, of the developing zygote have the capacity to produce any other cell in the body, but they also have the genetic instructions for producing the extraembryonic cells that form the placenta in the uterine wall, which is necessary for the zygote to fully develop.

Some commentators have discontinued using the term "totipotent," as explained by Jonathan Slack: "The word 'pluripotent' has now become the standard term for describing the ability of ES [embryonic stem] cells to form a broad range of cell types. ES cells used to be described as 'totipotent,' but this usage was discontinued because mouse ES cells do not normally form the outer layers of the embryo called the trophectoderm when allowed to differentiate *in vitro*."[3]

Other cells found in the early embryo approximately four days into its development (at which point it is called a blastocyst), while less powerful than the most primitive totipotent embryonic stem cells, can form all of the approximately 200 cell types found in the human body. These cells are called pluripotent. A third variety of stem cells, called multipotent, have already undergone differentiation and can produce a limited number of tissue cells. These are also sometimes referred to as adult stem cells. Some adult stem cells referred to as unipotent (i.e., spermatogonial) can only form a single cell type (i.e., sperm).

■ ■ ■

Scene: Charles Walker is a stem cell biologist, one of the nation's pioneers in isolating, activating, and delivering embryonic stem cells to tissues in rodents, and a recipient of the prestigious Lasker Basic Medical Research Award. After receiving her master's degree in bioethics and before she started her own career in stem cell research, Dr. Franklin interviewed Dr. Walker in his laboratory for an article she was planning to write for a general audience, including people like her father.

FRANKLIN: When did your interest in stem cell research take off?

WALKER: As you know, James Thomson, at the University of Wisconsin, isolated and cultured cells from the inner cell mass of early human embryos. These became the first human embryonic stem cell lines. His work generated enormous interest in human stem cell research. A year after that, I became interested in the research.

FRANKLIN: Was this the first time you'd thought about working with stem cells?

WALKER: I had a course in cell biology that focused on stem cells, but it was mostly devoted to blood stem cells. I was interested in solid tissue, so I did not follow it up.

FRANKLIN: What were the blood stem cells used to treat?

WALKER: For more than forty years, blood-forming or hematopoietic stem cells have been used in bone marrow transplants to treat leukemia. But there were still no stem cell lines. Then in 1981, 1988, and 1995, stem cell lines were created from mice, hamsters, and primates respectively. Like most progress in science, the discoveries in stem cells occurred in small, incremental steps.

FRANKLIN: I know the different types of stem cells. By what criteria do scientists distinguish them within the new nomenclature?

WALKER: Stem cells are classified by their developmental potency or plasticity, that is, the range of other kinds of cells they *can* become—or, put another way, their ability to differentiate into other cells. The greater their potency, the more differentiated cells they are capable of producing. Totipotent cells can form all the cell types in the body, plus the extraembryonic, or placental, cells. Embryonic cells within the first couple of cell divisions after fertilization are the only cells that are totipotent. Pluripotent cells can give rise to all the cell types that make up the body, and adult stem cells are multipotent, which means they can produce several types of cells in a tissue.

FRANKLIN: Is there something in the genetic structure of the cell that distinguishes it as totipotent, pluripotent, or multipotent?

WALKER: Probably not. We identify stem cells from a tissue or the blood based on their functionality. They need to be able to reproduce themselves continuously and differentiate into different cell types.

FRANKLIN: Is it correct then that the development of the organism progresses from totipotent to pluripotent to multipotent, then unipotent stem cells, and then to a specific cell type?

WALKER: That is essentially correct. But the totipotent and pluripotent cells last for a very short time in embryonic development. They become differentiated and disappear after about a week, which is why they can be harvested from a very early embryo. A unipotent stem cell can only produce one lineage of cells, and so has the narrowest potential for differentiation. For the purpose of cell regeneration, the multipotent and unipotent or adult stem cells are more prevalent and closer to the differentiated tissue cell type, but they do not have the plasticity or developmental potency of embryonic stem cells. Thus it is not as easy to get them to replicate in sufficient quantities and to differentiate to the cell of choice.

FRANKLIN: If stem cells, whether totipotent, pluripotent, multipotent, or unipotent, are potentially helpful to replace diseased, damaged, or abnormal cells, what has to happen?

WALKER: We must have cells that proliferate extensively—that is, they need to be robust. They need to differentiate to the desired cell type for replacement. They need to survive in the patient after a transplant; they need to integrate seamlessly into the surrounding tissue. And they need to function in the recipient's body as healthy cells during the recipient's life. Finally, they need to be safe and not tumorigenic, and cannot cause any rejection or immunological effects.[4]

FRANKLIN: This suggests that there is something stem cells have that the other cells don't have. Is that correct? Is there something in the stem cell that defines it as such?

WALKER: The issue is not totally resolved. It may be that the stem cells, especially embryonic stem cells, have all the switches for development open, whereas when the cells begin to specialize, or differentiate, some switches on the DNA are closed. So the defining element is not in the DNA but in the epigenome—the proteins that surround the DNA, or chemical modifications to it.[5]

FRANKLIN: How do you unleash the power of the stem cell to do the things you would like it to do?

WALKER: By and large, it's about signaling. You have to understand and apply the appropriate biochemical signals. It is like knowing the combination of a lock to enter a vault. Some scientists argue that there is nothing intrinsic to a cell that makes it a stem cell, and it's all in the signaling. One scientific group wrote: "Which cell becomes the stem cell is thus determined by the feedback, and not anything intrinsic to the cell."[6]

FRANKLIN: How do you know the right signal to give to the stem cell? Is there a dictionary or cookbook of stem cell signals?

WALKER: The manual of stem cells is not complete. There is a lot of trial and error. We put human embryonic stem cells into dishes that have lots of little wells in them—kind of like Bingo cards with three or four hundred little wells in each. Next we treat each well with a different chemical, and we look for which chemicals tell the cell to move one step further. In this systematic way we are able to move forward step by step in directing the differentiation of cells into the type of cell we would like.[7]

FRANKLIN: This sounds like a *lot* of trial and error.

WALKER: It starts out with a lot of trial and error. Nevertheless, a great deal is already known about what chemical signals turn stem cells into different tissue and bones, which are very similar but not identical to the natural substances.

FRANKLIN: Okay, so it's the signaling. But scientists have been able to turn an adult cell into a stem cell by a kind of reprogramming, like turning back the clock on the cell. Is that all about signaling too?

WALKER: The cells you are talking about are called induced pluripotent stem cells. They were first created in 2006 by the Japanese scientist Shinya Yamanaka, who was awarded the 2012 Nobel Prize for Physiology or Medicine for converting mature cells to stem cells. He did it by adding four genes to skin cells from a mouse. Within a few weeks, the skin cells were reprogrammed into pluripotent stem cells—the precursor cells that were capable of forming (differentiating into) the skin cells. In this case the reprogramming involved a kind of genetic engineering, which is more than just signaling the cell. Perhaps the proteins coded by these genes were needed to turn on the genetic switches that were turned off as the pluripotent cell differentiated into the skin cell.

FRANKLIN: Thus far you have spoken about the remarkable potential of stem cells, but you haven't said anything about the risks. How likely is it that something as medically powerful as stem cell technology would come risk free?

WALKER: I am assuming you mean the risks to the well-being of the patient, not the moral risks of working with embryos.

FRANKLIN: Yes. Let's focus on the health risks when the stem cells are used to treat human illnesses.

WALKER: Bone marrow transplants involving hematopoietic stem cells have been safely done since the late 1960s, but there are risks of rejection. Donor matches and the new generation of antirejection drugs have significantly lowered the risks.

Stem cell lines developed in the laboratory from embryos or cord blood and transplanted into patients introduce a different set of risks. For example, a stem cell transplant was administered to a young boy suffering from a rare genetic disease called AT (ataxia-telangiectasia). The symptoms of the disease are poor body coordination, respiratory problems, and a weak immune system. Four years after the boy underwent fetal stem cell therapy, he developed tumors in his brain and spinal cord.[8]

FRANKLIN: Four years is a long time after the transplant. How do the doctors know there is a connection?

WALKER: They analyzed the tumor and discovered it did not match the boy's tissues. It came from a foreign source of DNA—the stem cells.

FRANKLIN: Was that an unusual case, or have there been other reports of tumorigenesis arising from stem cell transplants? And won't that limit the use of the therapy?

WALKER: It is not an isolated case. There have been other articles published about the risks of producing tumors from embryonic stem cells. Some report tumors in animal studies and some report tumors in clinical trials. For example, in the case of using human embryonic stem cells (hESCs) for spinal cord injury, scientists have identified two kinds of tumors in rat models: large benign teratomas produced by hESC-derived neurons and benign neural tissue–specific tumors.[9] Until the source of the tumors is determined and eliminated, it will not be used in human beings in this country—as far as I know.

FRANKLIN: Do these tumors appear shortly after the patient receives the transplant?

WALKER: Sometimes the tumors from the hESCs don't appear until years or decades later.

FRANKLIN: Is it fair to say that, in terms of risk, the more potent the stem cell is—the earlier it is in its stage of differentiation—the less predictable it is in what it can become?

WALKER: Yes! There seems to be an inverse relationship between potency and predictability. It is analogous to Heisenberg's Uncertainty Principle in physics, where the more precisely you measure the position of an elementary particle, the less precisely you'll be able to measure its velocity. The behavior of hESCs, which are at the earliest stage of maturation, is less predictable than that of stem cells from an advanced lineage, like adult stem cells. The early lineage cells can be more easily affected by where they are placed in the organism and the signals of the local environment.

FRANKLIN: Suppose medical scientists wish to use cultures of differentiated stem cells for therapy. Is it possible that those cultures still have undifferentiated stem cells or iPS cells? Wouldn't that present a problem?

WALKER: For sure. Transplanting to a patient even a small number of undifferentiated stem cells carries a risk of tumorigenesis. Scientists have to find a way to screen out the potentially dangerous cells.[10] You can prevent tumor formation by selecting the differentiated from the undifferentiated potentially tumorigenic cells. Another finding is that iPS cells may be more tumorigenic than embryonic stem cells, which makes iPS cell safety a high-priority problem.[11]

FRANKLIN: Are there other risks in transplanting hESCs, besides tumors and rejection?

WALKER: There's the risk that the transplanted cells will differentiate into the desired somatic cells but behave too aggressively in the organism. Imagine, for example, if islet stem cells were transplanted into a diabetic patient and started overproducing insulin.

FRANKLIN: Do you have an example?

WALKER: There was a report of two patients who received stem cell transplants for Parkinson's symptoms. The patients developed a disabling condition called dyskinesia—diminished voluntary movements and the presence of involuntary movements. Eventually it was learned from brain scans that the transplanted cells (neurons) produced too much dopamine. The patients were fortunate that there was an antidote to the aggressive dopamine-producing cells.[12]

Also, stem cell implants in China for chronic spinal cord injury resulted in some patients contracting meningitis. There have also been reports of nervous system complications when stem cells were used to treat blood diseases.[13]

FRANKLIN: So like any new medical technology, if anything good comes out of it, there will also be casualties on the path to success. But the history of stem cell research has a component not found in most medical innovations, and that is embryos—the richest source of pluripotent stem cells. There are continued debates about the safety of human subjects, and these are overlaid on debates about the moral status of embryos. As you know, stem cell research has become politicized—so much, in fact, that President Bush established a permissible embryonic stem cell line, a requirement if scientists wanted to receive public funds. This is unique in history.

WALKER: That is neither a defensible nor a sustainable science policy. The government cannot put limits on legitimate scientific inquiry.

Dr. Franklin's next visit is to an advisor of President George W. Bush to discuss the stem cell policy of the Executive Branch.

DIALOGUE 3

THE PRESIDENT'S STEM CELLS

In the wake of *Roe v. Wade* (1973), public sentiment turned to concern that aborted fetuses would be used for research. In response, Congress established a moratorium on the use of federal dollars for fetal research until a national commission could be established to sort out the ethical issues. The commission issued its report in 1975, lifting the moratorium on federal funds but setting strict guidelines on fetal research that applied to tissues derived from aborted fetuses. Research on live embryos awaited the establishment of an Ethics Advisory Board (EAB), which in 1979 recommended that IVF research on embryos up to fourteen days old was ethically acceptable.[1] Federal funding for early embryo research was then blocked by Congress. The newly elected administration of Ronald Reagan did not accept the EAB recommendations and failed to reappoint the board. Without it, there could be no federal funds. In 1990 George Bush Sr. vetoed a bill to override the ban on embryo research.

President Clinton issued an executive order in 1993 instructing his Secretary of Health and Human Services to end the congressional ban. Both an internal National Institutes of Health (NIH) Human Embryo Research Panel and President Clinton's ethics advisors recommended, based on the human health potential of the research, that it should commence using donated IVF embryos and aborted fetal tissue. However, faced with an intense backlash from anti-abortion and pro-life advocates, the Clinton administration failed to provide funding. Beginning in 1995 and continuing each year thereafter, Congress inserted a rider into its appropriations bill that prohibited the expenditure of federal dollars for creating or destroying human embryos (the Dickey-Wicker Amendment).

President George W. Bush entered office in January 2000. The following dialogue is a fictionalized account of a conversation between two fictional characters and is not intended to reflect the views of any real person. Any similarities between the advisor to George W. Bush and any actual person is purely coincidental.

■ ■ ■

Scene: The White House. Bernard Stein, M.D., is an ethics advisor to President George W. Bush, head of a national bioethics think tank, and a leading scholar on reproductive ethics. Dr. Franklin obtained an appointment with Dr. Stein to discuss President Bush's policies on human embryonic stem cells.

FRANKLIN: [*To Dr. Stein*] Thank you for inviting me to your office. As you know from our correspondence, I am an editor of the *Journal of Bioethics and Medicine*, and we are preparing a special issue on stem cells. Dr. Stein, can we begin by you helping me understand how U.S. policy on stem cells evolved? Did it arise in the Bush administration?

STEIN: The federal policy on human embryos was catalyzed largely after two events: first, the Supreme Court decision on abortion in 1973 and the first baby (Louise Brown in England) born after in vitro fertilization in 1978. After the *Roe v. Wade* decision, which made early stage abortions legal, a moratorium was placed on government funding for embryo research. Then in 1979 an Ethics Advisory Board to the U.S. Department of Health, Education and Welfare issued a report on the ethics of research involving human embryos. This advisory board said it was ethically acceptable to do research on embryos used for IVF purposes but postponed any recommendations on research involving the collection and culture of early human embryos fertilized naturally—not used for IVF. But they had one major caveat: the embryos could not be sustained in vitro beyond fourteen days after fertilization.[2]

FRANKLIN: Why did they set the boundary at fourteen days? That sounds quite arbitrary.

STEIN: At the fourteenth day of its development, an embryo exhibits a "primitive streak"—a faint white trace that is the first evidence of the embryonic axis. It is a precursor of the neural tube and the nervous system. Without a neural tube, there is no spinal cord, and the embryo cannot have feelings or exhibit any level of consciousness.

FRANKLIN: So the primitive streak is some kind of Maginot Line for bioethicists and shouldn't be crossed.

STEIN: In 1979 the hope was that establishing a moral boundary would allow scientists to continue with their embryo research, as long as they stayed within that limit.

FRANKLIN: Between 1979 and 1980 there was a change in administration. Jimmy Carter had lost the election to Ronald Reagan. Were the advisory board's recommendations adopted?

STEIN: Hardly. By 1980, the charter of the advisory board ran out and was not renewed. As you point out, Ronald Reagan was elected president. He and his administration opposed any research on embryos of any age. Republicans were, on the whole, more critical of research involving embryos than Democrats. But there were many Democrats who supported the moratorium.

FRANKLIN: Dr. Stein, let me see if I get this. The Supreme Court ruled that embryos are not persons, and therefore abortion was not murder, and established a fundamental right of women over their bodies, at least for the first trimester of pregnancy. And then a president opposed any federal funding for embryo research on the grounds that embryos could not be harmed. Why didn't Congress get into the act?

STEIN: Well, Congress did act, but not until another advisory committee was convened. In 1994, during the administration of President Bill Clinton, a federally appointed nineteen-member Human Embryo Research Panel issued its report. The panel concluded that embryos do deserve some moral consideration, but do not have the same moral status as persons because they lack specific capacities such as consciousness, reasoning, and sentience—at least, early embryos. The panel approved the use of federal funds for research on early embryos under specific guidelines.

FRANKLIN: Did that clinch it for President Clinton? After all, he is a Democrat and not doctrinaire on the issue. So he must have been receptive.

STEIN: No, it didn't work out that way. In 1994 NIH convened a Human Embryo Research Panel to draft guidelines on the use of federal funding for research on human embryos. The panel recommended that funding for creating embryos for research be permitted.[3] Clinton disagreed, but he was personally in favor of funding for scientific studies of embryos left over from IVF procedures. Nevertheless, responding to the political climate, Clinton wanted more deliberation and chose not to allocate federal funds to support research on leftover embryos until he could get a recommendation from a presidential ethics advisory committee. Perhaps he was anticipating congressional action.

FRANKLIN: Well, did Congress act then?

STEIN: Soon after the president made his preliminary decision to withhold funds, Congress closed the door on any research involving the destruction of a human embryo. The Dickey-Wicker Amendment (sponsored by Representative Jay Dickey, House Republican from Arkansas, and Roger Wicker, Senate Republican from Mississippi), which Clinton signed into law, has been attached to appropriations bills every year, starting in 1996. It essentially prohibits the Department of Health and Human Services from using appropriated funds for the creation of human embryos for research purposes or for research in which human embryos are destroyed.[4]

FRANKLIN: It seems to me there could be ways around the amendment. Suppose private money is used to create and destroy embryos and public funds are used to experiment on the cells removed from them. In many countries, like Germany, when a moral decision on embryo research is reached, it applies to everyone, not only those receiving funds from the government.

STEIN: You put your finger on one of the peculiarities of the bifurcated system of ethics in our country—one set of principles for public funding and another for private funding. Embryo ethics straddles two moral universes, and scientists have had to navigate through that thorny divide. They must establish a firewall between publicly funded and privately funded laboratories.

FRANKLIN: What happened with the president's advisory committee?

STEIN: In 1999, the president's National Bioethics Advisory Commission recommended that use of embryonic stem cells harvested from embryos discarded after in vitro fertility treatments—but not from embryos created expressly for experimentation—be eligible for federal funding. The Clinton administration decided that it would be permissible under the Dickey Amendment to fund human embryonic stem cell research, as long as such research did not directly cause the destruction of an embryo. In other words, women could donate their unused frozen embryos to research if the stem cell derivation, which involved destroying an embryo, was conducted under private auspices. Clinton published the guidelines for embryonic stem cell research on August 23, 2000. They allowed scientists to use federal funds to obtain stem cells from private suppliers who extracted the cells from donated frozen embryos.

FRANKLIN: But then President George W. Bush took office. As I recall, he embraced the Dickey-Wicker Amendment. It seems like he was opposed to using public funds for any stem cell research. It was, I believe, part of his moral/religious compass. Did he follow Clinton's plan?

STEIN: Hardly! Bush changed the playing field. He created his own Maginot Line. On August 9, 2001, before any funding was granted under these guidelines, Bush announced modifications to allow public funds to be used only for what was then a well-defined number of existing stem cell lines that were produced before a certain date—more than sixty embryonic stem cell lines that already existed from privately funded research. These were embryos on which the "life and death decision" had already been made.

FRANKLIN: Wasn't this number somewhat of an exaggeration, according to most scientists, since many of the stem cell lines were unusable? The cell lines eligible for NIH funding have been shown to have genetic instabilities. NIH-funded scientists would also like to have access to cell lines that have been derived without the use of animal feeder cells—a layer of mouse cells were typically used to culture human embryonic stem cells to activate their growth—or animal products that contaminate the human cells. Also, many cell lines that have been generated since that policy was put into place have mutations specific to certain human diseases like Huntington's, ALS, and Parkinson's that would be valuable for research into the progression of these diseases and for drug testing.[5] Other human ESCs exhibited chromosomal abnormalities that make them unsuitable for research. It strikes me that President

Bush was trying to appease his political base rather than listening to the scientific community.

STEIN: There were many complaints from scientists after the president's message. They claimed they could only use a minimum of eleven and a maximum of twenty cell lines that the Bush administration approved.

FRANKLIN: What you are saying is that President Bush had chosen a very restricted collection of stem cell lines—the president's stem cells—that were permissible to use under his watch.

STEIN: Essentially, that is correct. His advisors, I among them, felt that because these lines were already established—the unethical act of destroying human embryo life had already occurred—it would be ethically correct to do good things with them. But destroying new embryos, whatever their origin, was not ethically justified, and we shouldn't weigh costs and benefits to determine the correct moral choice.

FRANKLIN: But aren't we losing the opportunity to make remarkable medical discoveries that will save or improve the quality of human life? Is embryo life worth that loss?

STEIN: We must be concerned about going down a road where the early stages of human life become a natural resource to be mined for other people's benefit. There are alternative paths that can be pursued toward the same ends without destroying an embryo's life.[6]

FRANKLIN: It seems that the Bush administration had reached a conclusion about the moral status of embryos, and the scientists opposed to it were not persuasive. They did have a way out, though: get private funds to do their work.

STEIN: That is currently the situation. But under President Bush public funds were denied for research on embryonic stem cells if those cells were obtained by the destruction of embryos, regardless of whether that was paid for by public or private funds.

Dr. Franklin continues her inquiry into the moral status of the embryo to understand the bioethical foundations of the administration's policy decision. She attends a U.S. Appeals Court hearing where plaintiffs challenge a liberalization of the Bush administration's stem cell policy.

DIALOGUE 4

THE DICKEY-WICKER ENIGMA

U.S. Representative Jay Dickey of Arkansas, who served in the House from 1993 to 2000, and U.S. Senator Roger Wicker, who served from 1995 to 2007, contributed a rider to an appropriations bill introduced in 1995 and signed into law by President Bill Clinton in 1996. The so-called Dickey-Wicker Amendment prohibits federal funds to be spent on research that involves the destruction of a human embryo. It has been added to appropriations bills for the Departments of Education, Labor, and Health and Human Services every year since its inception.

The history of U.S. policy on embryo research can be traced back to the Supreme Court decision on abortion, *Roe v. Wade* (1973). Public fears that aborted embryos would be used in research brought attention to setting guidelines for and limits on the use of embryos in scientific studies funded by federal dollars. Under the Federal Policy for the Protection of Human Subjects, enacted in 1977, no Department of Health and Human Services (HHS) funds would be awarded for research involving embryos until an Ethics Advisory Board (EAB) was established and issued recommendations. This happened two years later, in 1979. The EAB issued recommendations for research on early embryos; however, President Carter took no action. Under the administration of President Ronald Reagan, the EAB's charter expired and was not renewed and thus, following federal law, no research on embryos could be publicly funded.

In 1993 President Bill Clinton signed the National Institutes of Health Revitalization Act (P.L. 103–43). Section 121(c) eliminated the requirement of an Ethics Advisory Board for federal funding to study human fertilization, giving the NIH authority to fund human embryo research.

The NIH established a panel of consultants including scientists, ethicists, and public policy experts to evaluate the ethical issues of working with embryos and to determine which embryologists and stem cell researchers should receive federal funding. The panel issued its report in 1994. The panel found acceptable certain areas of preimplantation human embryo research within a framework of strict guidelines.[1] The recommendations included permitting the use of spare embryos from fertility clinics for obtaining embryonic stem cells.[2] President Clinton did not accept all the panel's recommendations; he directed the NIH to prohibit the use of any funds for experiments that would create new embryos for research but was favorable to funding research on IVF excess embryos without appropriate consent provisions. In response to the 1993 act, the NIH panel's recommendations, and the president's policies, Dickey and Wicker introduced their rider to the appropriations bills.

The Dickey-Wicker Amendment stipulates that funds allocated under the Appropriations Act may not be used for the creation of human embryos for research purposes or for research in which a human embryo or embryos are destroyed, discarded, or knowingly subjected to risk of injury or death greater than that allowed for research on fetuses in utero under the Public Health Service Act (42 U.S.C. 289g(b)). The term "embryo" was defined as any organism, not protected as a human subject under the Code of Federal Regulations (45 CFR 46), derived by fertilization, parthenogenesis, cloning, or any other means from one or more human gametes or human diploid cells.

When President Bush took office in January 2001, he reviewed federal policies on embryo research and ordered DHHS to stop all funding for research involving embryos until it was studied further. In August, he issued a policy restricting federal funding for embryo research to a limited number of stem cell lines that had been created by that date. President Barack Obama took office in January 2009; in March he issued an executive order that overturned the Bush policy and expanded federal funding for research using stem cells obtained from excess embryos in IVF clinics. Two adult stem cell researchers, James Sherley and Theresa Deisher, sued the federal government over the new Obama policy on human embryonic stem cell (hESC) research. They argued that Obama's 2009 revision of the Bush rules violated the Dickey-Wicker Amendment. Chief Judge Royce Lamberth issued an injunction against the NIH funding of hESC research until he could fully consider the issues. The Obama administration appealed Lamberth's decision with the D.C. Court of Appeals.

The following dialogue provides a fictional account of the arguments heard by the U.S. Court of Appeals for the District of Columbia. It is based loosely on court documents, including primary and amici briefs, and does not purport to represent the actual statements made by the parties or the court. All characters in the dialogue are fictional.

■ ■ ■

Scene: Arguments are heard before the three-judge panel of the U.S. Court of Appeals for the District of Columbia regarding *Sherley v. Sebelius*—a challenge to the executive order by President Barack Obama allowing federal funds to be used for human embryonic stem cell research, as long as it uses only excess IVF embryos. The chief judge presiding is Leonard Bentley; the first circuit judge is Benjamin Anderson; the second circuit judge is Thomas White. The plaintiff's attorney with the Coalition for the Advancement of Medical Research is Phillip Barron; the defense attorney representing the administration is Assistant U.S Attorney Stephanie North.

JUDGE BENTLEY: The court will hear the arguments of *Sherley v. Sebelius.* In August 2009 the appellants and their amici filed a complaint against the Secretary of Health and Human Services and the National Institutes of Health, seeking relief from the NIH guidelines authorizing the funding of research involving embryonic stem cells. Will the plaintiff's attorney, Phillip Barron, present the argument to the court.

BARRON: Beginning in 1996 and carried through every year to the present, Congress has approved a rider to the federal appropriations bill called the Dickey-Wicker Amendment, which states that no federal funds can be used to create human embryos for research and that human embryos cannot be destroyed, discarded, or knowingly subjected to a risk of injury or death greater than that allowed for research on fetuses in utero. President Obama issued an executive order that permits federal funding for research on embryonic stem cells, which are isolated by destroying embryos. We believe that this executive order violates the Dickey-Wicker Amendment and that the congressional act would take precedent over the executive order.

JUDGE BENTLEY: Mr. Barron, the executive order does not state that federal funds can be used to destroy embryos. It only permits federal funds to be used to study embryonic stem cells that may have been produced legally by foreign researchers in their own country or by private companies that may have destroyed embryos to produce human embryonic stem cells without violation of federal law. Given that the executive order neither advocates nor explicitly permits the use of federal funds in violation of Dickey-Wicker, how do you reach your conclusion?

BARRON: Your honor, we believe it was the intent of the Dickey-Wicker Amendment to prevent the government's complicity in embryo destruction. Because research on embryonic stem cells is predicated on destroying human embryos, federal support for this research promotes the destruction of embryos, even if it is done by individuals or institutions that do not receive federal funds. Government funding for research on human embryonic stem cells, or hESCs, is enabling the destruction of embryos and is thus in direct conflict with Dickey-Wicker. The Dickey-Wicker ban unambiguously extends to *any* research project that uses hESCs. If a funded research project involves the use of hESCs, then embryos have been destroyed and Dickey-Wicker has been violated.

JUDGE ANDERSON: Mr. Barron, the exact wording of the Dickey-Wicker Amendment states that public funds shall not be used for "research in which a human embryo or embryos are destroyed, or knowingly subjected to risk of injury or death greater than that allowed for research on fetuses in utero." How do you interpret "in which" in this wording of the law?

BARRON: We interpret the term "in which" to mean embryos destroyed by an investigator with or without public funds. In other words, we believe that an NIH-funded investigator who acquired the embryonic stem cells from a private source who destroyed an embryo violates the law.

JUDGE ANDERSON: Mr. Barron, how do you interpret the term "research" in Dickey-Wicker?

BARRON: We interpret "research" in the act as an extended process that includes the initial derivation of human embryonic stem cells.

JUDGE BENTLEY: Ms. North, what is your response to the plaintiff's argument?

NORTH: Your honor, the administration has carefully promulgated an executive order, followed by the passage of guidelines by the National Institutes of Health. Both these documents are consistent with the provisions of Dickey-Wicker. Following the 1999 opinion of the NIH General Counsel, human embryonic stem cells are not human embryos within the statutory language of Dickey-Wicker. A human embryo as defined in Dickey-Wicker is an organism that is capable of becoming a human being when it is implanted in a uterus. Human embryonic stem cells are not such organisms because they are not capable of becoming human beings.

JUDGE BENTLEY: Ms. North, does your conception of "research" in Dickey-Wicker include the acquisition of human stem cells from embryos?

NORTH: The administration interprets "research" in the more traditional sense, as a "discrete project with starting materials, a method, an intervention, and an observed outcome." "Research" in this sense does not include the origin of the starting materials. As we read Dickey-Wicker, it permits federal funding of research projects that utilize already derived hESCs, which are not themselves embryos.

JUDGE BENTLEY: Mr. Barron, do you have any other arguments from the plaintiff's bar?

BARRON: Yes, your honor. We would like to affirm that even if the NIH guidelines do not directly violate the Dickey-Wicker ban on funding research in which human embryos are destroyed, conducting federally funded research on hESCs increases the demand for more embryos to create these cell lines, thus placing more embryos at risk as defined in Dickey-Wicker.

JUDGE WHITE: Mr. Barron, there are thousands of embryos kept in cryogenic tanks for IVF purposes that will never be used for human reproduction. If this is the major source of embryos for research and they already exist, how does the government policy increase the demand for new embryos?

BARRON: New eggs are more desirable for research than older frozen eggs. Frozen embryos certainly meet some of the demand, but there will still be increased demand for new eggs, especially if federal funding for research creates a commercial market for them, which it most assuredly will. The government's policy will incentivize future destruction of embryos.

NORTH: It is the government's view that the language of the Dickey-Wicker Amendment does not ban funding for research that provides an incentive to harm, destroy, or place at risk human embryos. Even if such effects were to occur, they represent unintended probabilistic effects of the statute, which lie beyond the legislative intent.

Sherley v. Sebelius was decided by the U.S. Court of Appeals on August 24, 2012. The three-judge panel ruled in favor of the defendant (the Obama administration). The decision included the following language:

> We held that NIH had reasonably interpreted Dickey-Wicker's ban on funding "research in which . . . embryos are destroyed" to allow federal funding of ESC [Embryonic Stem Cell] research. . . . We explained that "research" as used in Dickey-Wicker was a "flexible" (i.e., ambiguous) term. . . . It could be understood as the plaintiffs construed the term—an "extended process" that would include the initial derivation of stem cells. Or "research" could take on NIH's narrow interpretation as a "discrete project" separate from derivation. Given that ambiguity, we deferred under *Chevron* to NIH's permissible construction of Dickey-Wicker: "research" as used in Dickey-Wicker may reasonably be understood to mean a "discrete endeavor" that excludes the initial derivation of ESCs. Under that interpretation, Dickey-Wicker permits federal funding of research projects that utilize already-derived ESCs—which are not themselves embryos—because no "human embryo or embryos are destroyed in such projects."[3]

The researchers, supported by the right-to-life group Jubilee Campaign's Law of Life Project and the faith-based Alliance Defending Freedom, petitioned the Supreme Court to take the case, based on two principal arguments. The first was that the D.C. Court of Appeals erred in maintaining that President Obama's 2009 executive order could excuse an agency from complying with the Administrative Procedures Act. The second was that a preliminary injunction ruling was not binding law of the case. Lawyers for the petitioners surmised that in order to meet the court's conclusions about the ambiguity of the language in Dickey-Wicker, Congress would have to amend the act by changing the phrase "research in which" to "research involving" and the phrase "are destroyed" to "are or have been destroyed." The Supreme Court refused the petition on January 7, 2013.[4]

DIALOGUE 5

THE MORAL STATUS OF EMBRYOS

By "the moral status of human embryos" we usually mean our obligations to the nascent forms of human life. Embryos themselves have no obligations. In humans, the implantation of the embryo in the uterus begins the approximately nine months of prenatal development. Different nations and cultures have adopted a range of obligations of both the individual and the state to the developing embryo. The issue has become more prominent in recent years with the debates over abortion and the development of new reproductive technologies, especially in vitro fertilization, cloning, and therapeutic uses of embryonic stem cells.

The ancient philosophers argued that a fetus was not formed until at least forty days after fertilization for a male and eighty days for a female. Aristotle believed that our obligation to protect embryos starts when the embryo experiences sensations, and that the female embryo develops more slowly than the male. In biblical times the embryo and fetus were considered property. If a person were found responsible for causing a miscarriage, he could be charged a fee for the loss of property incurred by the mother.

The moral status of the embryo was connected with personhood in the Middle Ages. "Quickening," fetal activity at about twenty weeks felt by the mother, was thought to be the the moment of ensoulment and the first communication between mother and fetus. As the biology of reproduction advanced, distinctions were made between fertilization and conception, which is defined as "the onset of pregnancy, marked by implantation of the blastocyst or fertilized ovum in the uterine wall." Fertilization, on the other hand, is neither a single moment nor an event but a process that begins when a sperm attaches to the zona pellucida, the clear coating that surrounds the oocyte (egg). It is currently believed that the sperm genome may initially take control of embryonic development (until the diploid genome combining the DNA of sperm and egg takes over).[1]

So the moral status of the embryo depends on some combination of scientific, cultural, and religious assertions. A panel moderated by Dr. Franklin investigates some of these perspectives as she continues exploring how to relate to stem cell research.

■ ■ ■

Scene: A panel discussion with Rabbi Jacob Goldman, Monsignor Patrick Callahan, and Mary Osborn on the moral status of early embryos. Osborn is a secular humanist and the author of the acclaimed book *Ethics Without God.*

FRANKLIN: Since 1998, stem cells have become the fastest growing field of new research in biology. Cell biologists have recalibrated their research agendas to become associated with stem cell development and its therapeutic possibilities. Because the richest source of pluripotent stem cells is embryos, the debate on the moral standing of embryos has exploded. Prior to this, most ethical discussions about embryos grew out of the abortion controversy. But because embryos are a potential reservoir of stem cells that can be used to save or improve lives, embryo ethics and politics have taken on a new life. My first question to the panel is: When does an embryo become an entity that deserves moral consideration, rather than, say, a cluster of cells? Let's start with Father Callahan.

CALLAHAN: The Church has a very clear position. The moment a human egg is fertilized by human sperm (the union of gametes) and becomes a zygote, it acquires moral standing and the status of "moral subject"—a living entity that can be harmed or benefited. This means that humans as individuals and our society as a whole have an obligation to protect that nascent human life. By law the embryo is not a full human subject, but according to most Catholic teachings, the fertilized egg in every stage of its development has full moral status commensurate with that of a person.[2] From the point of conception there exists a human subject with a well-defined identity, its own coordinated, continuous, and gradual development, such that at no later stage can it be considered a simple mass of cells.[3]

FRANKLIN: Do you make a distinction between fertilization and conception? After all, there may be a few days before the finalization of the exchange and divisions of the chromosomes between the sperm and the egg, after which the embryo is prepared to attach to the uterine wall.

CALLAHAN: While there is no papal policy on the time of ensoulment, most Catholics accept that it is the moment of fertilization of the egg. For practical purposes, the Church sets the time of ensoulment as when the sperm enters the egg. The process of meiosis and the rearrangement of male and female chromosomes does not have to be completed. The critical factor is that life has begun. That's when God's will is expressed.

FRANKLIN: This very early form of life, just several days after fertilization, may be used to help cure or save the life of a fully formed child. By extracting the inner cell mass of the blastocyst, medical scientists can turn the pluripotent stem cells into heart tissue or brain tissue, both of which are desperately needed. Even if the zygote has some moral status, might that be preempted by the use it could be put to—to save a fully formed human being who is loved and part of a family?

CALLAHAN: The ablation of the inner cell mass of the blastocyst damages the human embryo. Such an act is gravely immoral. We do not distinguish between embryos that are wanted and those that are unwanted; in God's eyes, all embryos are wanted. Moreover, with regard to exploiting a nascent human life to help a fully developed human life, a good end does not make right an action that in itself is morally wrong. Extracting stem cells from a four-day-old blastocyst is like yanking organs from a baby to save other people's lives.[4] Just as we cannot accept having a baby in order to use its body parts, we cannot condone the use of an embryo for its parts.

FRANKLIN: Rabbi Goldman, how does your tradition address the moral status of the embryo? How much overlap is there in your views with those of Father Callahan?

GOLDMAN: I should preface my remarks by acknowledging that there is no single authoritative position within Judaism on embryos. There are several notable authoritative positions that probably reflect the range of what we find across society. I shall speak from my own tradition within Conservative Judaism—the progressive end. The moral standing of the embryo depends on two important factors: the age of the embryo and the sociomedical context—whether it is in the body or out of the body.

An embryo in the womb that is anticipated, loved, and considered part of a welcoming family has a higher moral standing than a similar age embryo that is in a freezer being held for a possible but not anticipated IVF treatment. Also, as the embryo develops from an eight-celled mass to a fetus with obvious human characteristics, it acquires a higher moral status.

FRANKLIN: Is your position consistent with a woman's right to an abortion?

GOLDMAN: My tradition holds that our bodies belong to God, and we have them on loan for the duration of our life. Our obligation is to preserve human life and health.

During the first forty days of gestation, the fetus, according to the Talmud, is "as if it were simply water," and from the forty-first day until birth it is "like the thigh of its mother." Within the forty-day period, an abortion can be used to save the life or health of the woman. Judaism recognizes psychiatric as well as physical factors in evaluating the potential threat that the fetus poses to the mother.[5] When abortion is prohibited in Jewish law, it is not because it is an act of murder, but as an act of self-injury.[6]

As the embryo gets older, the influence of the sociomedical context on its moral standing rises. Within the first forty days, the authority of the moral voice of the pregnant mother plays the strongest role. As the embryo ages, the role and moral responsibility of the community (and the state) become greater.

We should be clear that the embryo at any stage has some moral standing.

FRANKLIN: Do you approve of harvesting embryos for stem cells?

GOLDMAN: That depends on whether embryonic stem cells can be used to preserve or protect fully formed human life. There must be a moral imperative that exceeds the imperative of respect for the very early embryo. I would object to using embryos simply for research purposes that have no therapeutic possibilities. Life can only be destroyed or exploited for life.

We should not kill animals indiscriminately or for hunting as a sport, but only if it is done to enhance life. For the exploitation of early human embryos, it must be shown that it is done to improve human life and that there are no other reasonable options to achieve the same ends.

Another factor to consider about the use of embryonic stem cells is whether they are extracted from the donated embryo to cure diseases and protect health or for enhancement. Any genetic technology that is used for enhancement must be approached with extreme caution because of the eugenics campaigns directed against Jews in the early and mid-twentieth century.

FRANKLIN: Ms. Osborn, how do you construct a humanistic ethics of stem cell research and therapy and the moral status of embryos?

OSBORN: The moral basis of a humanistic perspective has two pillars. The first is scientific knowledge and the second is the enhancement of human life. Moral claims are not fixed as timeless truths but can change with advances in science and cultural evolution.

FRANKLIN: How does science affect the moral status of the embryo?

OSBORN: For one thing, it shows that we do not become individuated at conception. At a very early stage of development, the embryo can divide into twins, a process called monozygotic twinning, or remain a single individual. Alternatively, a pair of embryos that could result in fraternal twins can fuse together into one embryo and develop into one person (a tetragamete chimera). Before that stage is reached, we do not have an identifiable individual. Until the stage of twinning or chimerism has passed, which is about two weeks, we cannot say that there is a separate and unique individual.

CALLAHAN: Whether the womb has one or two individuals doesn't change the moral status of the embryo; it is still unethical to destroy life. I do not see the relevance of twinning or chimerism to the moral value of the embryo. Twinning simply means we have two human life forms to protect. The fact that a fertilized egg divides spontaneously does not mean that the initial egg or the two eggs from the mitosis lose any of their moral worth.

OSBORN: The point I am making is that the earliest stage of embryo development does not meet the conditions of there being a single entity. The early cells in the zygote are contiguous within a single extracellular membrane, the zona pellucida, but that does not make them a single entity, any more than placing a number of marbles in a sack turns them into a single entity. At this early stage, the cells are not yet functioning in an integrated way. There is no sentience, no consciousness, and, for embryos that are not attached to the uterine wall, no potential to develop into a human being.

FRANKLIN: Are there factors other than individuality that are relevant to the moral status of the embryo?

OSBORN: The development of the individual is one factor. The other important factor is sentience, namely whether the embryo can have feeling. And that cannot happen until the "primitive streak" develops. The primitive streak is a groove along the midline of the embryonic disk that establishes the embryo's cranio-caudal (head-to-tail) and left-right axis. It is this band of cells from which the integrated embryo begins to develop. Before the appearance of the primitive streak the embryo is a bundle of cells, not yet a unified structure. There can be no neural tissue before the appearance of the primitive streak, and therefore no nervous system and no possibility of sentience. The embryo cannot be considered a moral subject before its fourteenth day of development because it is not yet fully individuated, is not an integrated entity, and has no neurological system.

GOLDMAN: Mary, you claim that the early embryo is not a unified life form before the appearance of the primitive streak and that it suddenly—on the fourteenth day or thereabouts—makes the transition from a bundle of cells to a unified organism. There are two things your analysis fails to account for. First, the development of the embryo is a continuous process. It does not experience a mysterious quantum jump with the visible appearance of the primitive streak. Second, there is elaborate communication or cross-talk among the cells of the early embryo. This intercellular communication makes the appearance of the primitive streak as well as other essential submicroscopic structural components of the developing embryo possible. The unified structure begins at fertilization, not two weeks later.

OSBORN: A living human organism is an entity with human genes composed of differentiated living parts that function together in an integrated way to sustain a single life, not itself as part of another living biological entity. The cells comprised by an embryo in the first two weeks following conception do not yet serve sufficiently different functions to be coordinated in the service of a single life.[7] During the first couple of weeks after conception, all that exists is a collection of qualitatively almost identical cells living within a single membrane. Ontologically there is no organism.[8] The early embryo unattached to a uterine wall is an ambiguous, nonindividuated entity and deserves no rights, and if any at all, the most minimal moral status.

CALLAHAN: Although Mary's conclusion sounds like it is based on science, in reality, when the level of differentiation and communication becomes sufficient for us to say "there is a human organism" is not a biological question, but a metaphysical question. I believe that citizens have to approach this through their choice of a belief system, which includes science and religion, that provides the greatest unity, coherence, value, and meaning to their life experience.[9]

OSBORN: My view is consistent with the findings of the high courts. The Supreme Court of Tennessee, in *Davis. v. Davis*, ruled that the pre-embryo deserves respect greater than that accorded to human tissue but not the respect accorded to actual persons, because it has not yet developed the features of personhood, is not yet established developmentally as an individual, and may never realize its biological potential.[10]

FRANKLIN: Thank you, panelists. This discussion has demonstrated that science needs bioethics and vice versa—bioethics needs science.

DIALOGUE 6

CREATING GOOD FROM IMMORAL ACTS

The history of Western ethics has been largely guided by two grand theories: deontology (represented by the ethics of Immanuel Kant) and utilitarianism (represented by the ethics of John Stuart Mill). Deontology is based on the fundamental rightness or wrongness of an action, whereas utilitarianism looks at the total balance of good versus evil when evaluating an act.

In the wake of heinous crimes committed by the Nazis during World War II, German science was critically examined and deemed responsible as an enabler, contributor, and participant. Jews, Gypsies, the mentally challenged, and other minorities living in Germany or German-occupied territories were forced to participate in unethical experiments, which have since been examined in books like *The Nazi Doctors* by Robert J. Lifton and *Racial Hygiene: Medicine Under the Nazis* by Robert Proctor.

Some members of the postwar scientific community believed that published works by German scientists who collaborated with the Nazi regime should never be used or cited in the scientific literature. Data published from unethical experiments, it was argued, could not be trusted, and scientists who participated in Nazi war crimes should not be honored by having their published work incorporated into the edifice of science. Besides, their work would never be cited or replicated, and thus was not reliable.

But what if some of these unethical experiments produced unique data, which no one else had, that could save lives? Is it ethical to create good out of immoral science? Stephen Post wrote: "Because the Nazi experiments on human beings were so appallingly unethical, it follows, prima facie, that the use of their results is unethical."[1] He argues that science must draw a line between civilization and the moral abyss (the *summum malum*) around which ethics builds fences.

The issue of Nazi science may be an extreme example in considering whether evil acts should be exploited for any good they can produce. In the debate about stem cells and the destruction of embryos, some scientists began to raise analogies with extreme cases. If living human embryos have some ethical status, even if not fully as persons, can destroying them be justified for the common good? Can one benefit from the results of what some believe to be a past immoral act without becoming complicit in the act?[2]

■ ■ ■

Scene: Dr. Franklin flies to Germany to discuss the politics and ethics of stem cells with scientist Gordon Baum. Franklin is interested in how German scientists view their responsibility to the law and ethical norms while they investigate the medical benefits of human embryonic stem cells. She wants to understand why Germany does not permit scientists to destroy embryos in order to derive human embryonic stem cells, but allows them to work with embryonic stem cells produced this way outside the country.

FRANKLIN: Dr. Baum, there are clearly sharp legal and moral divisions over the use of human embryonic stem cells. Some countries, like Germany and the United States, have prohibitions against destroying human embryos for research, including as source materials for embryonic stem cells. Other countries, like the United Kingdom, allow research on and destruction of embryos prior to fifteen weeks. You live in a country where it is illegal to destroy human embryos to obtain embryonic stem cells. If that is considered illegal and possibly immoral, why would it be legal or moral to work with stem cells taken from embryos destroyed in countries other than Germany?

BAUM: Germany, my country, has a reprehensible history of eugenics during the dark period of the Nazi regime. This regime selected certain people and certain fetuses as undesirable. So as a nation we are extremely sensitive about selecting or destroying embryos. Other countries are not burdened with this historical legacy and so are not as morally vigilant. Germany has chosen to create moral buffer zones to prevent the society from getting too close to embryo selection or destruction. So we can feel comfortable using hESCs derived by other countries because our experiments do not involve embryo research, selection, or destruction.

FRANKLIN: Let us assume, for the purpose of our discussion, that destroying embryos is wrong. If you as a scientist could use the products of destroyed embryos for good purposes, would you consider that ethical? Is it wrong to create good from immoral acts committed by others?[3]

BAUM: I guess there are two issues underlying your question. Can we do good things through immoral acts? And should we use the outcome of bad acts to do good things? The answer to the first question is most assuredly yes. We tell lies or violate a law (for example, pass a red light) if we know it will save someone's life. Utilitarian ethics dictates the ethical outcome of such a choice. Ethicists use many examples to illustrate how, by violating a moral norm to protect another person, we are doing the morally correct thing. The acts in question are usually specific in time and place, not continuous activities, such as telling a lie to an intruder to protect a family member. But when there is a continuous activity that violates legal or moral rules, the issue becomes more complex, and utilitarian ethics might not produce the best outcome.

FRANKLIN: Can you be specific?

BAUM: A continuous act would be something like torture. It is possible to justify a single act of torture on utilitarian grounds, based on the number of lives saved. But if we were to generalize the use of torture, it would clearly be immoral.

FRANKLIN: Aren't you using Kant's categorical imperative?

BAUM: Precisely! It is not possible to consistently universalize a maxim such as: "I can use torture whenever I believe the act will produce more good than evil."

FRANKLIN: Your second question was "Should we use the outcome of bad acts to do good things?".

BAUM: We need to look at the cases. For example, I have a physicist friend in the United States who had cancer as a young man in the 1950s. Not much was known about shielding the body from radiation therapy. German scientists under the Nazi regime did immoral experiments and produced the only data available at the time for determining the amount of lead shielding needed to protect the testes during radiation therapy. My friend used the results of these immoral experiments to advise physicians on the proper radiation shield, and he was able to have a healthy child after his treatment.

FRANKLIN: Many people believe it is wrong to buy products made by exploited child laborers. Suppose a nonprofit organization that houses, feeds, and clothes the homeless had an opportunity to purchase a crate of clothing at a very low price, but the clothes were manufactured by child labor. Would it be ethical to purchase them and distribute them to poor families?

BAUM: The critical question is whether you become an enabler of an immoral workplace. Your purchase of those products, no matter what good they do, provides support for an immoral and illegal operation. The outcome might be different if the clothes known to have been manufactured with child labor were found in an abandoned warehouse. In that case you could argue that your action to distribute the clothes would not support the illegal activities.

FRANKLIN: Now we are left with the question: If there are no universal norms or international laws preventing the destruction of embryos, only laws in distinct nation-states, is it immoral for you to use imported human embryonic stem cells that have been acquired from destroyed embryos in another country?

BAUM: The answers may differ depending on whether you are viewing the question prospectively or retrospectively. Jonas Salk's polio vaccine was tested on mentally retarded orphans without reasonable informed consent. We would not permit these experiments today. But no reasonable person would refuse the benefits of polio vaccine based on the ethical transgressions that took place in the 1950s and 1960s, before we had regulations on informed consent in medical experiments.[4]

FRANKLIN: Looking prospectively, if killing embryos is immoral, then is it ethically justifiable for scientists to use human embryonic stem cells from a black market source for research designed to find the cause of and treatment for diseases?

BAUM: Let me first respond by saying that I object to the word "killing" applied to a few cells—a zygote—so early in development that they exhibit no human qualities. If there were an international consensus on embryo creation and destruction for research, then I would be the first to say that it would be immoral to accept embryonic stem cells on the black market. Without an international convention, I do not consider it immoral to use the products of fertilized human eggs a few weeks into their development, when obtained according to legal and ethical procedures in a country that allows stem cells to be sent to other countries, even countries that prohibit destroying embryos.

FRANKLIN: Let us suppose there was an international treaty that prohibited the production and destruction of embryos for research. Some have argued that using stem cells from embryos does not constitute an endorsement of their destruction. As an analogy, imagine that environmentally insensitive strip miners are engaged in practices that unnecessarily kill hundreds of fish. If nearby residents eat the fish instead of letting them rot, the residents are not endorsing the fish-killing practices. They are simply finding a positive outcome from an ecological tragedy.

BAUM: Again, we need to distinguish cases that are one-of-a-kind from those that describe continuous events. We also have to acknowledge those cases where people who acquire some benefit from an immoral act are explicitly or implicitly reinforcing the continuation of the act, where they become enablers. The high demand for embryonic stem cells, like the high demand for illicit drugs, reinforces the production of those cells and therefore is a tacit endorsement of embryo destruction. The fish story does not help us through the moral quagmire. Research on donated embryonic stem cells, harvesting embryonic stem cells, and destroying embryos in order to obtain stem cells are all part of the same enterprise. The researcher is not like a person who searches the Dumpster and finds clothes made by underage workers. We cannot avoid the core ethical problem—whether it is always immoral to destroy a human embryo—by hiding behind the supply chain.

FRANKLIN: The IVF industry produces tens of thousands of excess embryos. Many will be destroyed when they are no longer needed. So why not use them for research?

BAUM: The reason for our policy in Germany is that our scientists do not want to participate in the supply chain of embryo destruction. But once the embryos are destroyed elsewhere, we will do good science with the stem cells. From a moral standpoint, for the pure researcher, the intention of those who destroyed the embryo is irrelevant.

FRANKLIN: In the United States, two sets of ethical guidelines prevail, one for federally funded research and another for privately funded research. And yet the government under the Bush administration did not allow public funding for studying human embryonic stem cells derived in the private sector. Does this seem like a consistent and viable moral position?

BAUM: The crux of the matter is whether you believe that, regardless of societal benefits, destroying an embryo created for research purposes is prima facie wrong, whether legal or not. If that is the accepted belief, then one cannot justify using human embryonic stem cells obtained in this way. Following this logic, if a scientist used human embryonic stem cells and saved a life, then the result would have to be evaluated by head-to-head comparison with other immoral acts committed to save a life.

FRANKLIN: What if an embryo is considered a person?

BAUM: Then the moral equation changes. It is universally accepted that one person cannot be sacrificed to save another. All civilized societies have embraced the principle of Immanuel Kant that a person is an "end-in-him/herself" and not a means to some other end. Personally, I cannot accept the idea that a zygote less than fifteen days old fulfills the concept of personhood.

FRANKLIN: There are over 100,000, some say as many as 400,000, human embryos stored under cryogenic conditions. The vast majority of these fertilized eggs, which are held for family building, will eventually not be used for IVF.[5] They will either be selectively destroyed or over time will no longer be viable. Is there any ethical reason these embryos should not be donated for research?

BAUM: If destroying an embryo is intrinsically immoral, then donating it to research will not redeem the act of its destruction. We donate our organs after death, but it would be immoral to kill a person before they die, even if their death is imminent, to harvest their organs. Because

I do not believe it is intrinsically immoral to harvest stem cells from a week-old embryo, I cannot accept the conclusions that follow from that premise.

FRANKLIN: Suppose someone believes that the moral worth of embryos is not absolute, but still believes it is immoral to destroy them for harvesting stem cells. Yet they take a different stance on the use of externally fertilized embryos for reproduction. They see the moral status of the embryo differently in the context of research and reproduction. If that is a defensible moral position, then after the cryogenically preserved embryos are released from their use in IVF and await destruction, why shouldn't they be used to harvest embryonic stem cells?

BAUM: This case seems as close as you can get to living tissue donation. Even those who make a distinction between embryos for research and embryos for reproduction seem to believe that "throwaway embryos" should be available, at the consent of the donor, for contributing to human good through research. And yet there are bioethicists who would disagree with me. A Catholic priest, asked by a parishioner what she should do with her excess IVF embryos, replied that she should either put them up for adoption or pay their rent (in the freezer) for life.

FRANKLIN: But the frozen embryos will eventually die.

BAUM: Yes, but under Catholic teachings, allowing an embryo to die a natural death is morally distinct from destroying it to harvest its stem cells. The outcome may be the same, but the intentions are different.

FRANKLIN: If our moral respect for embryos is really a surrogate for the respect we are supposed to have for people, using doomed embryos for research seems less likely to diminish respect for human life than simply routinely discarding them in fertility clinics. Almost any way of putting embryos to use in research sends a much better message than just discarding them.[6]

DIALOGUE 7

CIRCUMVENTING EMBRYOCIDE

After President George W. Bush prohibited the use of public funds for embryo research, there remained deep divisions in public opinion on the issue of destroying embryos for harvesting embryonic stem cells. Some scientists became uncomfortable in their complicity with embryocide. In seeking alternatives, they considered working on deactivated embryos, in which critical genes had been removed to render the embryo incapable of implanting in the uterine wall. They believed that research on embryos that had no potential to become a human life would address public concerns. Nevertheless, there remained opposition to embryo research leading to destruction, whether the embryo was disabled or not. The potential of an embryo to become a person has been used to oppose embryonic stem cell research, but negation of that potential was not used to support the research.

Then a remarkable discovery in reprogramming adult cells to return to a stem cell state resulted in a sea change in the debates. After several years of experiments on mice, two scientists reported their reprogramming of human somatic cells in 2007. Working independently, James Thomson[1] at the University of Wisconsin and Shinya Yamanaka at Kyoto University reprogrammed human fibroblast cells into pluripotent stem cells.[2] Their processes essentially reversed the development of the adult cells and brought them back to an embryonic-like state. Yamanaka used four genes to obtain the reprogrammed human cells. Following their success, in November 2012 researchers from Austria, Hong Kong, and China published a protocol for generating what were called human induced pluripotent stem cells (iPSCs) from exfoliated renal epithelial cells present in urine.[3]

Some scientists who began their research using embryonic stem cells switched to iPSCs so they could apply for federal funds. The headline in *The New York Times* read: "Scientists Bypass Need for Embryos to Get Stem Cells."[4]

Would induced pluripotent stem cells make the controversy over embryonic stem cells moot? And would iPSCs introduce any unique risks when used on humans? The pace of discovery was rapid: in April 2009 it was demonstrated that iPS cells can be created without the use of any genetic elements or any genetic alteration of the adult cell. This was achieved by a treatment of certain proteins channeled into the cells.[5] The term "protein-induced pluripotent stem cells" (piPSCs) came into use.[6] Protein-induced pluripotent stem cells, however, are not easily reproducible, and using modified RNA to induce pluripotent stem cells has advantages over the use of proteins.[7] A young Japanese scientist, Haruko Obokata, was the lead author of a 2014 *Nature* publication that claimed she had reprogrammed somatic cells into iPSCs by lowering the pH (or raising the acidity), thereby stressing the cells.[8] According to many observers, this procedure, called stimulus-triggered activation of pluripotency (STAP), was the simplest method to date for creating iPSCs from mature cells. When problems were found in the study and scientists failed to replicate it, the results came under investigation by the respected RIKEN Center for Developmental Biology in Kobe, Japan.[9] A dozen other research groups reported that they could not replicate the results. On April 1, 2014, Riken found Obokata guilty of misconduct.[10] Three months later, on July 2, *Nature* retracted two papers it had published about research involving STAP.[11]

■ ■ ■

Scene: When Dr. Franklin learned of the new research producing iPSCs, she reached out to a prominent critic of embryonic stem cells, Catholic bioethicist Paul Flannery, to discuss the ethical implications of the new findings and the alternatives to destroying embryos to acquire pluripotent stem cells. Flannery has been advocating against the use of human embryos for harvesting stem cells. Dr. Franklin challenges his view and proposes keeping all research opportunities open to scientists, a policy she refers to as "stem cell pluralism."

FRANKLIN: Dr. Flannery, I am delighted to meet you, having read a body of your work on bioethics while working on my degree at U Penn. I want to start by hearing your analysis of the debate over embryonic stem cells.

FLANNERY: People's opposition to destroying human embryos to acquire stem cells has deepened even while their interest in the benefits of stem cell technology in the treatment of disease and replacing damaged tissue has increased. As a result, some scientists have pursued other avenues that may obviate the need to destroy a viable embryo. I believe this approach will get us out of the moral abyss that characterizes our current situation. What is your take on this?

FRANKLIN: I agree! New approaches to stem cell development beyond the use of embryos are emerging. As you know, one of the most promising is to create induced pluripotent stem cells, or iPSCs.

FLANNERY: I haven't kept up with this new technique. How is this done? What do you start with?

FRANKLIN: It is an incredibly fast-moving field of research. Anything I say today could be moot tomorrow. The first iPSCs were produced in 2006 from mouse cells, and a year later, more were produced from human cells. The process typically starts with a somatic cell like a skin cell. To reprogram it, several techniques can be used. The first time this was done for human cells, James Thomson at the University of Wisconsin and Shinya Yamanaka at Kyoto University isolated four genes that were found to be important to embryonic stem cells and transferred them into adult cells. The resulting cells exhibited the properties of pluripotency. The four genes used by the two groups were not identical.

FLANNERY: How did they get the genes into the adult skin cells?

FRANKLIN: Typically, they use viruses. Viruses are great vectors because they have evolved to infect cells and deposit their genetic baggage; this is called viral transfection.

FLANNERY: So these genes enter and reprogram the adult stem cell. Can they do anything else?

FRANKLIN: Several of the genes used for reprogramming adult skin cells are oncogenes—tumor-causing genes. If the reprogrammed cells with oncogenes are implanted into someone, they might cause cancer.

FLANNERY: So you are saying that scientists can reprogram adult cells into pluripotent stem cells without having to create or destroy embryos, but that these cells can cause cancer if introduced into humans? What's the point? Has anything been gained?

FRANKLIN: Oh, a lot has been gained. It's a proof of concept. Once you know that you can achieve the reprogramming, you refine the process so that there is no risk that the reprogrammed cells will induce tumors.[12] Another group, at Harvard University and led by Konrad Hochedlinger, used a different virus—an adeno virus—to transport four genes into adult cells. The adeno virus does not deposit any of its own genes into the adult cell, which reduces the chance of creating tumors.

FLANNERY: You said there were other methods of reprogramming cells.

FRANKLIN: There are some promising results. Yamanaka has been able to accomplish reprogramming with circular pieces of DNA called plasmids—without using viruses. And quite remarkably, in 2009, scientists at the University of Edinburgh and the University of Toronto reported that they had created iPS cells by inserting proteins into adult cells. Can you foresee any ethical issues?

FLANNERY: The pursuit of human iPS cells raises new challenges for the process of informed consent. People who donate their cells for iPSC research must provide voluntary and informed consent. Researchers will want to collect cells from individuals suffering from serious medical conditions to be studied as an iPS cell line. The donors must be informed about how the resulting iPS cells made from their cells will be used. Remember how Henrietta Lacks was treated to obtain her valuable immortal HeLa cells? It is no longer ethical to acquire human cells without the informed consent of the donor.

FRANKLIN: What uses may be problematic to donors?

FLANNERY: Some donors may be uncomfortable with the idea that their cells may be used to develop human-animal chimeras or hybrids. (A chimera is an organism comprised of cells from two or more different species. Scientists can learn things about human genes by transporting them into the genome of animals.) It may be against people's religious or ethical beliefs to have their cells used in this way.[13] Also, some may resent the idea of scientists making a profit from the exploitation of their cells.

FRANKLIN: The creation of chimeras has benefited science and medicine. Human brain cells have been transplanted into the mouse genome for the study of human neurological diseases and chimeric mice with human immune cells have been used in testing drugs for AIDs. Of course, anyone participating in a clinical trial must approve the use of their cells to develop an iPS cell line. But there are many banks of stored tissue. To derive iPS lines from these tissue banks, no informed consent is required.

FLANNERY: This is a gaping hole in our regulatory system. When people donated their tissue to the tissue bank, they didn't know about iPSCs and couldn't fathom that their cells would be used to make a human-animal hybrid.

FRANKLIN: It could be extremely expensive and difficult to obtain post hoc informed consent for the use of tissue samples deposited years or even decades ago. The samples in many of these tissues are numbered and anonymized. It may not even be possible to track down the donor. Restrictions on the samples would be a great loss to science.

FLANNERY: There is another ethical issue concerning iPSCs. Now we know skin cells can be driven back to a pluripotent state. But imagine that this is not the end. Let us suppose the skin cells can be driven back even further, to a totipotent state. In such a case, any cell in a person's body could be reprogrammed into a totipotent cell and thus could be induced to produce a cloned human being.

FRANKLIN: Don't you think we have enough real ethical issues on our plate before we get into the hypothetical ones? So far, no one has reprogrammed an adult cell back to its totipotent state. And if that did happen, we would have to address the issue of human cloning, which arises also from other methods that have been successful in animals. A more plausible technique for acquiring embryonic-like stem cells without destroying the embryo has already been accomplished. Scientists can select one cell from the inner cell mass of the pre-blastocyst stage without destroying the embryo. That single cell is pluripotent and can be used to develop a stem cell line.

FLANNERY: You may not be destroying the embryo, but you are disabling it. Those cells in the blastocyst are not like Leibniz's monads, totally isolated from one another. They may be interacting in ways we still do not understand. Pulling out one cell could be damaging to the embryo. So it's not a neutral intervention. The cells in the zona pellucida of the blastocyst make the embryo a single organism, not simply a cluster of cells. The fact that nutrients flow from outside to inside the zona pellucida and within different cells inside the membrane makes the multicelled organism a human life and therefore deserving of moral consideration. There is only one way of circumventing embryocide, and that is to prohibit it.

FRANKLIN: Okay, let me offer another technique. Suppose scientists create an ersatz human embryo that shares some of the properties of a human embryo but can never develop into a fetus. Scientists have known about parthenogenesis for at least a century. This is a form of asexual reproduction where animal eggs can be induced to develop into an embryo without fertilization—sometimes referred to as the "virgin birth embryo." This is usually done by an electrical or chemical stimulus of the egg. Parthenogenesis on human eggs, called parthenotes, has been accomplished, and these nonfertilized eggs have been brought to the blastocyst stage.[14] In 2006 Italian scientists did this, and from their parthenotes they were able to isolate pluripotent stem cells from the blastocysts and induce them to produce neurons.[15] A similar result of deriving human embryonic stem cells from parthenogenic blastocysts was accomplished by Chinese scientists in 2007.[16] This ersatz human embryo can be a source of embryonic stem cells but can never be used to create a person, so we are not killing a real human embryo.

FLANNERY: From all the science I know, parthenogenesis is a method of creating artificial life and circumventing the most fundamental natural process, the union of sperm and egg. But induced parthenogenesis in mice and monkeys frequently results in abnormal development. The idea of developing stem cells from abnormal development and then using those cells to correct a medical condition is absurd. As far as the ethics of creating an ersatz blastocyst, it's like manufacturing defective human embryos. It's a little like engineering human (or quasi-human) life. With parthenogenesis, you certainly avoid creating a natural human embryo, but the process violates other ethical intuitions about creating subhuman creatures.

FRANKLIN: We agree that the parthenote is not a natural human embryo. Titanium is not a natural human body part, yet it is used for artificial hip replacement. Why shouldn't we be able to use artificial embryos to cure people's diseases?

FLANNERY: Not so fast. The question is: Does a parthenote have any moral significance? Stimulating an unfertilized egg with electrical signals so it divides a few times is doing research on the egg. The egg does not attain moral status from the electrical stimulation. Nevertheless, some ethicists regard the parthenogenetically fertilized egg as a "nonviable embryo" that undergoes its first cell divisions normally and therefore deserves some moral status.

FRANKLIN: Okay, I will give you another example of obtaining stem cells from embryos without destroying them. Robert Lanza, chief scientific officer of Advanced Cell Technology, created two hESC lines using individual cells taken from three-day-old embryos made up of eight cells each. An embryo can lose one cell and still develop into a person. This happens all the time with preimplantation genetic diagnosis (PGD). Strictly speaking, we can obtain embryonic stem cells without destroying the embryo.

FLANNERY: Preimplantation genetic diagnosis has certainly become an accepted practice to enhance IVF reproduction—although the Catholic Church officially opposes IVF, and PGD along with it. There is still an intentionality distinction between modifying a human embryo for research and doing it for reproduction. If the scientist removes one of the eight cells to create an embryonic stem cell, what will he do with the remaining seven-cell blastocyst? No one will adopt it. Eventually it will be destroyed. Thus, the technology egg can be used to extract a cell without destroying an embryo during the IVF procedure. But for research purposes, the embryo will not be saved.

FRANKLIN: If I understand you correctly, from your moral standpoint, the least objectionable approach to avoiding embryocide is creating iPSCs where an embryonic-like stem cell is created from a somatic cell—a kind of "virgin birth pluripotent stem cell."[17]

FLANNERY: Yes, we like the idea.

DIALOGUE 8

MY PERSONALIZED BETA CELLS FOR DIABETES

Diabetes is a group of disorders characterized by persistent high blood sugar levels (hyperglycemia). The two most prevalent kinds of diabetes are type 1, or juvenile diabetes, and type 2, which affects adults, usually later in life. The common feature of these disorders is the body's inability to produce a sufficient amount of the protein hormone insulin, essential for regulating the body's glucose concentration in circulating blood. Insulin is produced by a group of cells called beta cells, located in the pancreas. In a normal individual, when the amount of glucose in the blood rises, the pancreas released more insulin to push more glucose into the cells, whereupon the blood glucose level drops. Without sufficient insulin, the blood glucose rises to life-threatening levels.

Type 1 diabetes, which affects about three million people in the United States, is an autoimmune disease, where the body's immune system attacks itself. The body's T-lymphocytes destroy the insulin-producing beta cells. The causes for type 2 diabetes are not well understood, although a number of environmental and genetic risk factors have been identified. Obesity and lack of exercise are two environmental risk factors. Approximately thirty-six diabetes susceptibility genes contributing to type 2 diabetes have been discovered.[1] Either the pancreas produces very little insulin or the body does not respond appropriately to the insulin. This is known as "insulin resistance."

The current treatment for type 1 diabetes consists of insulin injections, transplanting the cluster of islet cells, which consist of four distinct cell types, including insulin-producing beta cells, or a whole pancreas transplant. The insulin is either extracted from cows or pigs or produced from human genes as human insulin (Humulin), using genetically engineered bacteria. As the science of stem cells advanced, scientists conjectured that they could use pluripotent stem cells to create "beta-like" cells that would produce insulin but would not be attacked by the body's immune system. A California bond referendum (Proposition 71) established the California Institute of Regenerative Medicine (CIRM), which moved into its San Francisco headquarters in 2005 and began funding projects in 2006. In 2009 the California startup company ViaCyte received one of the largest awards ($20 million) from CIRM for its R&D work on diabetes treatment. The research awards to the company accumulated to $40 million in subsequent years. ViaCyte received FDA approval in August 2014 for the first human trial to test the safety and efficacy of its stem cell-derived replacement therapy for diabetes.[2]

■ ■ ■

Scene: Beverly Simpson has been living with type 1 diabetes her entire life. Her endocrinologist, Janet Richfield, is a stem cell scientist working on producing islet cells that can be transplanted to diabetics to cure their insulin deficiency. Dr. Richfield is discussing with her patient the progress in clinical trials involving personalized embryonic stem cells that eventually could lead to a cure for her diabetes. Dr. Franklin received permission to participate in their discussion.

RICHFIELD: Beverly, it's great to see you. Dr. Rebecca Franklin is joining us. She is a research medical scientist and physician working in areas that may help cure diabetes someday. How are you doing? How is your blood sugar control?

SIMPSON: I find that I have to check my blood more frequently; I seem to be having more trouble keeping the blood sugar in balance. And my diet is getting narrower and narrower.

RICHFIELD: As you know, diabetes can change as one ages. We have to calibrate your diet and your insulin pump. But there is something else I would like to discuss with you. The new field of stem cell research offers some promising results in treatment for type 1 diabetes. The idea is that the stem cell can be turned into an islet cell that produces insulin. Once these cells are transplanted safely into your body, you will begin producing insulin naturally.

SIMPSON: Where do these cells come from? How does this work?

RICHFIELD: I shall defer to Dr. Franklin.

FRANKLIN: The great promise of this process is that the cells come from you. In other words, we extract a mature cell from your body, say a skin cell. The nucleus of the somatic, or skin cell is then inserted into an enucleated egg cell—an egg from which the nucleus has been removed. The somatic cell nucleus is reprogrammed by the host cell. The egg is then stimulated with an electric shock, and it begins to divide. After many divisions in culture, this single cell forms a blastocyst—an early stage embryo with about a hundred cells—with almost your identical nuclear DNA.

SIMPSON: So now you have my DNA in cells that came from a human egg contributed by some anonymous donor. How are these cells going to help me?

FRANKLIN: After the human egg with your DNA begins dividing, the cluster of cells it generates are embryonic-like stem cells. The next step in this process is to make these stem cells differentiate into the beta cells (within the cluster of islet cells) that produce insulin.

SIMPSON: Why go through all this trouble? There are islet cell transplants available when the donors match.

FRANKLIN: In this case, you are the donor. And because it is your DNA, there will not be a problem with tissue rejection, which can happen even if there is a predictably good match, because the DNA is not identical. We are talking about producing a stem cell that can be turned into insulin-producing cells that are personalized for you, to be transplanted into your pancreas. The process is called somatic cell nuclear transfer, or SCNT.

SIMPSON: How far along is this research?

FRANKLIN: Work on pancreatic islet transplantation has been going on for a few years. But SCNT is fairly new and was first done to develop human embryonic stem cells in 2013.[3] A journalist named Alex O'Meara wrote a book titled *Chasing Medical Miracles*, describing his own experience in a clinical trial involving islet cell transplantation.[4] At the end he gives advice to anyone who is thinking about going into a clinical trial.

SIMPSON: I guess I should read his book. What was his experience? How did he do?

FRANKLIN: O'Meara received islet cell transplants from an external donor and had to take antirejection medication. His islet transplants were not sustainable, so he could not be weaned from his insulin injections. Whole pancreas or islet cell transplantation is available to a limited number of patients. It also requires long-term immunosuppressive therapy. But science has advanced, and the SCNT method is considered the most promising approach for curing diabetes. However, it is not near the advanced clinical trial stage yet in the United States.**SIMPSON:** I've been researching this on the Internet. There are a number of clinics outside of the United States, in Germany, Mexico, and China. Some are advertising treatments for diabetes using SCNT. I'm not interested in waiting another five or ten years. If stem cells might help cure diabetes, I want to be the first to benefit.

RICHFIELD: SCNT may be the most promising approach, but considerable hurdles must be overcome before it is deemed safe. Because diabetes is treatable, the risks to patients in a clinical trial cannot be too great. SCNT is potentially safer than pancreatic islet transplantation from matched donors, but it has risks that have not been evaluated. Currently, over 100 active clinical trials have been completed or are actively seeking to produce islet cells that can be transplanted into diabetics.

SIMPSON: Why would there be risks if they are only transplanting my cells back into me?

RICHFIELD: The process selects one of your adult cells and uses its DNA—46 chromosomes. But your DNA has been around for a while, exposed to radiation and chemical mutagens. Scientists can't be sure that the mutations will be expressed in the embryonic stem cells. Experience has taught us that most embryos created by SCNT are malformed. The method was used to create Dolly the sheep, and that took 277 attempts. Then Dolly exhibited many pathologies, even though the embryos had been pre-screened.[5] She was created by reproductive cloning using SCNT, and therapeutic cloning is clearly a different process; however, they share some of the same risks. And since there are already some partial successes with pancreatic islet transplants, the SCNT program has to be proven more effective.

SIMPSON: But islet cell transplants are happening today. The process you're describing is not even in use yet, and the risks are totally unknown.

RICHFIELD: Based on the results we have thus far, the islet donor transplants may only be a short-term fix because they won't reproduce themselves for the life of the patient. There is some disagreement about whether our bodies' stem cells will regenerate insulin-producing beta cells.[6]

SIMPSON: I've learned that there is a shortage of islets for transplantation. Postmortem pancreas donation provides islet cells only for a small percentage of the diabetics who could benefit from a transplant. Maybe the shortage could be solved by the new SCNT technology, once it's developed and approved. And the immune-suppressive drugs can make patients vulnerable to infections, but the drugs are unnecessary with SCNT. Is that right?

FRANKLIN: Yes, that's the potential. The issues you raise are precisely those that have spawned new therapeutic approaches to diabetes. Animal studies were successful in transplanting bone marrow-derived stem cells into diabetic mice. The investigators found that the stem cell treatment reduced hyperglycemia within seven days and kept it down until the mice were sacrificed at thirty-five days after transplantation. Based on the success of the animal studies, a clinical trial began in 2009, transfusing autologous umbilical cord blood into twenty-three children in an effort to regenerate pancreatic islet insulin-producing beta cells and improve blood glucose control. These studies are based on the assumption that with the right signaling, the stem cell-derived islet cells will be able to colonize the pancreas and continuously produce insulin.[7]

RICHFIELD: One of the problems with using SCNT to make embryonic stem cells is the availability of eggs. Over 200 eggs can be used to make one viable stem cell line. Dozens of women would be needed to generate the eggs. The U.S. National Academies of Science has come out against the commercialization of eggs. They do not believe that egg donors should receive financial inducements.[8]

SIMPSON: Of course, I am young enough to use my own eggs, although in a few years I might have none left. Maybe I should start harvesting my eggs now and freeze them until the procedure is approved.

RICHFIELD: But your eggs may have the DNA mutation that brought you diabetes in the first place. So it would not be wise to develop embryonic stem cells from your own eggs.

SIMPSON: So I need to purchase or get donation eggs so that scientists can establish the stem cells for my transplant. I wonder if they are selling embryos on EggAuction.com?

RICHFIELD: Two other problems have to be solved before autologous stem cell transplants can be done. There is no guarantee there will not be rejection of autologous transplants, because the mitochondrial DNA from the foreign egg will be different. Even if it were autologous, it could cause an immune reaction. Second, individual autologous products are very expensive and too difficult to make for each patient. In the future, we're going to have to work with an off-the-shelf product that is not completely matched.[9]

SIMPSON: What else is on the horizon?

RICHFIELD: A company called Viacyte has created a device called Encaptra—an artificial pancreas. Beta cells are enclosed in a porous and synthetic envelope, smaller than a business card, that is inserted just under the skin. The envelope allows blood to flow in and insulin to flow out.

SIMPSON: Has it been tested on humans?

RICHFIELD: Thus far, it has been tested in mice and has managed the blood sugar levels of animals with experimentally induced diabetes.

FRANKLIN: First we have to make it work for the best case situation; then we must deal with the cost.

SIMPSON: I hope I am a good candidate for beta cell transplants when the procedure is ready for clinical trials. In the meantime, I may ask my nondiabetic sister to freeze a few of her eggs. At least we have the same mitochondria.

DIALOGUE 9

REPAIRING BRAIN CELLS IN STROKE VICTIMS

Annually, about 800,000 Americans and 120,000 people in the UK suffer a stroke. About a quarter of them will suffer a second stroke. Stroke is the largest cause of disability and the third most common cause of death (after heart disease and cancer); the annual cost to treat victims in the United States is estimated by the Centers for Disease Control at $37 billion, including the cost of health care services, medications, and missed days of work.[1] Stroke can produce paralysis, permanent brain damage, cognitive impairment, loss of speech, and long-term disability. The first stroke raises the risk of a second and increases the risk of mortality. About 87 percent of strokes are classified as ischemic, meaning that a blocked artery interrupts the flow of blood and thus oxygen to the brain. Each minute there is blood loss to the brain, an estimated 2 million brain cells die. Those cells cannot regenerate by themselves. Scientists have developed strategies to regenerate damaged brain cells with stem cells.

Studies have demonstrated positive outcomes in animal models of ischemic strokes treated with implantation of stem cell-derived cells, where the stem cells were obtained from fetal or embryonic tissue, bone marrow, or peripheral and umbilical cord blood.[2] Currently the only FDA-approved treatment for ischemic stroke is tissue plasminogen activator (tPA), a protein that breaks down blood clots.[3] tPA is a serine protease (a family of enzymes that cut certain peptide bonds in other proteins) found on endothelial cells, the cells that line the blood vessels. As an enzyme, it catalyzes the conversion of plasminogen, a blood protein that is a precursor to plasmin, the major enzyme responsible for clot breakdown.

Fetal tissue has been the major source of stem cells in animal models of stroke treatment. Because of the potential risks of using human embryonic stem cells to repair brain tissue—such as teratomas, tumors composed of tissue foreign to the site of growth, and highly malignant teratocarcinomas, malignant teratomas found most commonly in the testes—there has not been a flood of research on transplanting ESCs in animal brain tissue. Induced pluripotent stem cells have not yet been studied in the treatment of strokes.[4] According to one review, the mechanism of cell replacement in stem cell therapy for stroke victims may play a less important role than originally believed. However, recent results in animals suggest that administering stem cells to animals after an induced stroke activates the outgrowth of nerve fibers of neurons in the brain that restore function.[5] If stem cells are expected to make an impact on regenerative medicine, successful treatment of stroke victims would go a long way toward that goal.

In the following dialogue, Dr. Leonard Hendricks is a fictional character who is not intended to resemble any real person. The statements attributed to him about ReNeuron are loosely based on the company's public comments about its activities, but do not purport to represent actual statements made by any person or representative of ReNeuron.

■ ■ ■

Scene: Dr. Franklin interviews a vascular surgeon and specialist in stroke treatment, Leonard Hendricks, from the United Kingdom. He is a consultant to the UK company ReNeuron, which has pioneered developing stem cell therapeutics for stroke victims. Dr. Franklin is trying to understand the relationship between stem cell development for stroke victims and spinal cord injury. Because she does not engage in clinical work, she hopes to learn about the challenges in moving from research to clinical applications.

FRANKLIN: Dr. Hendricks, you collaborated on the first clinical trial using human fetal stem cells in a stroke victim, called the PISCES study—Pilot Investigation of Stem Cells in Stroke. This was reported to be the world's first regulated clinical trial of neural stem cell therapy for stroke victims.[6] What prompted you to participate?

HENDRICKS: According to the World Health Organization, fifteen million people suffer stroke worldwide each year. Of these, five million die and another five million are permanently disabled. Nearly three-quarters of all strokes occur in people over the age of sixty-five, and the risk of having a stroke more than doubles each decade after the age of fifty-five. So, stroke is the most feared affliction of the elderly. It can cause impairments of speech, of cognition, and of movement due to damage to brain cells from lack of blood. Until recently, there hasn't been any hope of reversing the effects of a stroke. Families are devastated both emotionally and financially in caring for the victims, many of whom have a significantly diminished quality of life. The goal of the stem cell treatment is to improve their mental and physical functioning.

FRANKLIN: The use of fetal stem cells for regeneration of damaged brain cells and tissues is unproven. Was there a proof of concept that motivated you and others to use fetal stem cells on stroke victims rather than on people suffering from other diseases, like Parkinson's?

HENDRICKS: There are no corroborating animal models of using fetal cells to cure or improve chronic stroke, and animal studies cannot be easily equated with each other or related to humans. There have been a number of medical breakthroughs that do not follow the traditional path of laboratory bench to animals and then to humans.

FRANKLIN: Animal studies involve artificially inducing a stroke—is that right?

HENDRICKS: Yes. Animals would ordinarily die outright before they have a stroke.

FRANKLIN: I hear you saying two things. First, animal models are not very useful in predicting therapies for human stroke victims. And second, we need animal models before we can go into clinical trials. How do you reconcile these?

HENDRICKS: We need animal models, even weak ones, to eliminate the outliers—the therapies that have no chance of succeeding. Then we are left with a smaller set of possibilities. We keep narrowing these down until we have an intervention that just might work.

FRANKLIN: Are you saying that successful rodent studies, for example, are useful even though the etiology of human strokes and induced rodent strokes are so different?

HENDRICKS: Many rodent studies have demonstrated that stem cell transplantation by direct injection to the brain or by intravenous infusion can improve stroke recovery.[7] Fetal tissue has been the major source of cells for transplantation into animals in which a stroke has been induced.[8]

FRANKLIN: Did you use embryonic stem cells to derive neuronal cells?

HENDRICKS: No! ReNeuron developed a multipotent stem cell line from nonembryonic human fetal brain tissue taken from the cortex region of the brain.

FRANKLIN: But how did you know what the risk would be for your trial subjects?

HENDRICKS: We estimated that the biggest risk in injecting human cells into a person is the risk of rejection and associated side effects—and those can be dangerous. So we had to be fairly confident that the cells we used would not produce a serious immune reaction.

FRANKLIN: How did you recruit patients for the trial?

HENDRICKS: We looked for male patients sixty years of age or older who were moderately or severely disabled after suffering an ischemic stroke.

FRANKLIN: How did you introduce the stem cell-derived neural cells? Were they implanted into the brain? Were the patients fully informed about the risk?

HENDRICKS: The treatment involved a single injection of one to four doses of our specialized stem cell line (ReN001) into the patient's brain through a neurosurgical operation during which they were under general anesthetic. The first two patients received 2 million cells, the next three 5 million, the third group 10 million, and the fourth 20 million. We used an extensive informed consent procedure that discussed potential risk. We had to get consent from a family member if the patient had lost the ability to talk or respond to questions.

FRANKLIN: How long were they monitored?

HENDRICKS: The patients were monitored for outcomes over twenty-four months—the first phase trial was to evaluate safety, so we were looking for any pathology from the introduction of the cells. At the same time we could be cognizant of any changes in the patients' abilities.

FRANKLIN: *Nature Biotechnology* wrote that a major issue about the treatments is the lack of knowledge concerning the exact mechanisms by which implanted cells ameliorate symptoms. Were you flying blind?

HENDRICKS: In any clinical trial, we can never be sure about the mechanisms even though we have some ideas from animal tests. We want the cells to begin repairing damage—durable engraftment, so that the holes in the brain will be filled by active tissue, not dead tissue. A female patient with Batten's disease—an inherited fatal disorder of the nervous system afflicting children, who begin losing their vision—had a transplant of male donor cells. After she died they found the male donor cells engrafted in the brain. This type of evidence gives us confidence.

FRANKLIN: When were Phase I tests completed?

HENDRICKS: By September 2011, the Independent Data Safety Monitoring Board reviewed the safety data from the stroke victims' stem cell therapy trials, including the data from neurological exams and neurofunctional tests. The positive results were no cell-related or immunological adverse events reported in the patients treated. In addition, there were some reductions in neurological impairment compared to the patients' pre-treatment baseline. The Monitoring Board recommended that the trial could proceed to Phase II, in which higher doses of the ReN001 stem cell line were used.

FRANKLIN: Is it difficult to get informed consent for such trials?

HENDRICKS: The probability of a successful intervention is greater for healthier patients, but we have a more difficult time getting consent from them. When the patient has extreme cognitive impairment from a first stroke, the family is more likely to give consent to an intervention.

FRANKLIN: Does the family typically agree to accept a trial where a placebo may be given in lieu of a therapeutic intervention?

HENDRICKS: Mostly not. They want hope for improvement. In some unusual cases we can forego the placebo.

FRANKLIN: But if you forego the placebo, how do you know that your stem cell intervention works? Perhaps there would be improvement without the stem cells.

HENDRICKS: Our ethical position is first and foremost, "Do no harm." If the intervention doesn't work, the patient is no worse off. At least the family can hold on to some hope.

FRANKLIN: Doesn't this place a high burden on the safety studies and the inference that if it is safe in animals, it will be safe in humans?

HENDRICKS: That is true. The healthier stroke victims have both the most to gain and the most to lose from an experimental intervention.

FRANKLIN: ReNeuron chose multipotent rather than pluripotent cells for the clinical trial. Did that introduce any limitations?

HENDRICKS: The pluripotent human embryonic stem cells and iPS cells have greater plasticity than the multipotent cells. Some scientists have argued that although the multipotent neural cells can differentiate into neurons and glia cells, they will not be able to rebuild all the functioning brain tissue that was lost from the stroke. According to one report, if the lost tissue is to be reconstructed, it will require pluripotent stem cells.[9] But that does not mean there will not be some benefits from this therapy.[10]

DIALOGUE 10

REVERSING MACULAR DEGENERATION

The human retina contains several types of cells, including photoreceptor cells, retinal ganglion cells, retinal pigment epithelium, and inner nuclear layer cells. Macular degeneration is a disease that affects the photoreceptor cells, causing vision loss in the center of the retina, called the macula; a certain amount of peripheral vision remains. The retina is part of the central nervous system. It is unable to regenerate neurons damaged by disease.[1]

There are two basic forms of macular degeneration, known as dry and wet. Dry macular degeneration is age related and affects 1 percent of people over age 50 and 10 percent of those over age 65. Dry age-related macular degeneration (AMD) is the leading cause of legal blindness in people 65 or older in the developed world. Currently it is estimated that 1.75 million people suffer from AMD in the United States and an additional 7 million are at risk. Worldwide, it is estimated that 14 million people (1 in 2,000) are blind or severely visually impaired because of AMD. Dry AMD accounts for 90 percent of the cases.

Wet macular degeneration occurs when blood vessels invade the retina (called choroidal neovascularization, or CNV), destroying the retinal pigment epithelium that supports the light-sensitive photoreceptor cells. The primary treatment involves injecting a drug (an angiostatin) in the eye that blocks the growth of new blood vessels.[2] The drug has to be given repeatedly to stop new vessels from invading the retina. The treatment does not reverse previous damage, however; at best, it stops new leakage into the macula.[3]

Unlike cold-blooded vertebrates, humans cannot regenerate retinal cells. In 2000, scientists at the University of Toronto found stem cells in the eyes of mice, cows, and humans.[4] Ophthalmic scientists began investigating whether the retinal cells could be regenerated with adult or embryonic stem cells. Promising experiments have been performed on pigs, retrieving progenitor cells from healthy eyes and transferring them into eyes lacking those cells.

Induced pluripotent stem cells have opened up new research strategies for repairing the damaged retina. In one planned study, skin cells would be taken from an AMD patient's upper arm and reprogrammed to make induced pluripotent stem cells, then turned into retinal cells. A small sheet of iPSC-derived retinal cells would be placed under the damaged area of the retina and induced to grow and repair the pigment epithelium.[5]

By 2013 there were thirteen clinical trials of treatment using stem cells for macular degeneration planned, in recruitment, or in progress listed on the U.S. Clinical Trials database (www.clinicaltrials.gov). Strategies include transplantation of already differentiated retinal cells, embryonic stem cell transplantation, and autologous transplants of induced pluripotent stem cells. In the next dialogue, Dr. Franklin sits in on a discussion between a stem cell scientist working on dry AMD and his mother, who has been diagnosed with the disease.

■ ■ ■

Scene: Ben Townsend is a stem cell scientist working with a team of ophthalmologists who has spent the last nine years trying to develop stem cells for the treatment of macular degeneration. His mother, Barbara Townsend, is a professor emeritus in the history of medicine at the University of Pennsylvania. She is in the early stages of macular degeneration and is preparing an article on stem cell therapy to treat the disease. She is also looking into entering a clinical trial. Dr. Franklin became acquainted with Professor Townsend at U Penn, and they have remained friends. Professor Townsend knew about Dr. Franklin's interests and invited her to a discussion.

BARBARA: Ben and Rebecca, would you care for a fresh-brewed cappuccino before we begin?

BEN: That would be great!

FRANKLIN: Yes, Barbara, that would be wonderful.

BEN: Mom, I know you are writing a journal article on your case and the role of stem cells in treating macular degeneration, but before we start, what was your last examination like?

BARBARA: Dr. Feldstein said I have dry age-related macular degeneration—dry AMD—and that it is progressing, but not very quickly. I can still read and write with my glasses, but I need a lot of light. Let's begin with you describing what is happening inside my eye.

BEN: As you know, our group is working on using stem cells to treat AMD as well as childhood macular degeneration. We began the first clinical trials for AMD on January 2011.

BARBARA: I know a little about AMD: that there is a sheath of cells in the central part of the retina that die and cannot repair themselves. These cells are critical for the eye to transmit signals to the optic nerve at the base of the retina.

BEN: That's a great beginning, but then it gets more complicated. There are two kinds of cells in the retina: retinal pigment epithelium (RPE) and photoreceptor cells. The death of both kinds is responsible for your loss of vision.[6] The photoreceptor cells cannot survive without healthy RPE cells, so both kinds need to be replaced by transplantation. Then the photoreceptor cells must be set in place.

BARBARA: Are they planning to replace both kinds of cells in a single operation, or is it a two-stage process?

BEN: It is not clear at this point whether an operation will be needed for each cell type or two kinds of cells can be delivered simultaneously in one procedure.

FRANKLIN: Have they succeeded in doing these experiments in animals?

BEN: Yes, Rebecca, there have been promising studies where transplanted embryonic stem-cell derived photoreceptors have integrated well into mouse retinas and restored light response.[7]

BARBARA: Will they be using embryonic stem cells for this in the human trials?

BEN: There has been some success in using iPS cells to improve the cells of the macula. They would take a skin cell from a patient and use it to develop a stem cell through reprogramming, then make RPE cells and photoreceptors from the iPS cells. In the clinical trial by Advance Cell

Technology (ACT), they used human embryonic stem cells to derive RPE cells in patients with Stargardt macular dystrophy, a juvenile form of macular degeneration.

BARBARA: How many children are affected?

BEN: This disease affects over 25,000 Americans and approximately one in 10,000 children.

BARBARA: How safe is the procedure?

BEN: ACT also undertook a clinical trial of the safety and tolerability of subretinal transplantation of human embryonic stem cell-derived RPE cells in patients with advanced dry age-related macular degeneration.

FRANKLIN: Wouldn't it be safer to use cells from the individual being treated—reprogram them into stem cells and then use them to make the cells of the macula—than to use a discarded embryo of another individual?

BEN: Most scientists believe that it is preferable to generate pluripotent stem cells from the patient's somatic tissue. With autologous grafts of various tissue types, we hope to be able to enhance or replace damaged cells in every organ of the body, including the eye.

BARBARA: Is there a concern about rejection with human embryonic cells from foreign embryos?

BEN: Yes, to some extent. Although the eye is quite resistant to these rejection effects, as what we call an immune-privileged site, it's not invulnerable. The consensus is that using autologous grafts is preferable to using replacement cells derived from embryonic stem cells.[8]

FRANKLIN: I have read that one of the risks associated with embryonic stem cells is that when they are transplanted they can develop into teratomas, or developmental tumors.

BEN: There have been cases like that in animal studies. To avoid this problem, scientists can either produce the corrective cells with human embryonic stem cells in a petri dish and transplant them into the eye or shut down the genes for teratomas in the embryonic stem cells before transplanting them.

BARBARA: Has there been much progress in using induced pluripotent stem cells to treat macular degeneration?

BEN: The Japanese scientist Masayo Takahashi and her team developed RPE cells in culture first from embryonic stem cells, then from iPS cells. Then they injected a thin sheet of the cells under the retina to replace the sheet of RPE defective cells in a clinical trial. Takahashi's team did this more than 100 times without tumor formation in animals. For each of the early patients, they used a tumorgenicity test for their iPS-RPE cells.[9]

BARBARA: Well, it's good that these cells can be administered tumor free, but does it work? Is it effective?

BEN: The jury is still out. The clinical trials have not been completed.

FRANKLIN: I read in *Scientific American* that scientists have used gene therapy to restore the vision of women who were born virtually blind.[10]If that is a proven method, shouldn't they use it for AMD?[11]

BEN: The gene therapy treatment was for a condition called Leber's congenital amaurosis (LCA), in which the photoreceptor cells in the retina have been destroyed. Those afflicted with the disease have a genetic mutation in the RPE cells that prevents them from making an enzyme that turns vitamin A into retinol. The photoreceptors need retinol to detect light and send signals to the brain.

BARBARA: So how did they cure the women with this inherited genetic disease?

BEN: They isolated the corrective genes from another individual, placed them in a vector, and injected the genes directly into the eyes of the patients. This therapy would not work for AMD, because that condition is not believed to be caused solely by a specific genetic mutation. We don't know the cause of AMD. So we have to replace the deteriorating cells without knowing why they have died.

BARBARA: What will stop the replacement cells from dying?

BEN: If they do, we hope it will take many years.

FRANKLIN: Aren't there some genes associated with AMD?

BEN: AMD probably results from a combination of genetic and environmental factors. Some genetic clusters are associated with increased risk of AMD, but scientists have not yet been able to pinpoint specific mutations.

BARBARA: So they have used human somatic gene therapy to treat AMD, but have they been able to treat it with stem cells?

BEN: Researchers treated two women with AMD with stem cells and got some results, but they are not sure it was from the stem cells.

BARBARA: So Ben, do you think I should enroll in the clinical trial?

BEN: Mom, you are still able to see. I would wait until we get results from the trial from patients who were almost 100 percent blind, so we can learn more about any risks.

BARBARA: Of course, my article would be more credible if I discussed my own trial.

DIALOGUE 11

MY STEM CELLS, MY CANCER

Cancer remains largely an enigmatic disease, according to some scientists—more accurately, a group of diseases with a family resemblance. Some tumors stay put, some metastasize. Some are hormone responsive, some are not. Theories of cancer etiology have evolved and include viruses, somatic genetic mutation, dysfunctional immune systems, and abnormal tissue environments. Cancer cells have an uncanny ability to survive treatment and re-occur. Genomic studies of cancer cells have turned away from organ site classification to the commonality of mutations among cancer cells. The cancer cell taxonomy may be revised from point of origin to type of mutation.[1] In 2013 oncologists proposed narrower rules to define cancer, introducing the neologism "incidentalomas," which they define as lesions or growths that will never cause a problem.[2]

Tumor cells exhibit a range of proliferative capacity, sometimes robust, at other times rather feeble. Scientists have doggedly investigated the biochemistry behind this capacity[3] and have advanced several hypotheses. The cancer stem cell hypothesis is one of the more recent that have caught the interest of cell biologists and oncologists.

According to this idea, there are somatic cancer cells (those usually identified by pathologists) and cancer stem cells. The latter possess the stem cell capacities of self-renewal and multilineage differentiation. "The definition of CSCs as the only self-renewable tumour cells capable of seeding a new tumour implies that CSCs are also responsible for initiation of metastases."[4] In rodent studies it has been found that only a subset of cells (between .0001 and 0.1 percent) have the capacity to develop into a new tumor.[5] Studies of immunodeficient mice show that cells that produce tumors are a biologically unique set with stem cell properties.[6]

From the hypothesis, it is inferred that malignant cells originate from a set of cancer stem cells.[7] If the growth potential of cancer depends on such a rare type of cells, then scientists want to know how to eradicate them, or at least prevent them from growing and reestablishing themselves at other sites.

Based on xenotransplanted human cancer cells in mice, only one in a million human melanoma cells is tumorigenic, although some scientists question whether this result understates the tumorigenicity of the cells:[8]

> The question of whether cells with tumorigenic potential are common or rare within human cancers has fundamental implications for therapy. If tumorigenic cells represent small minority populations, as supported by evidence supporting the cancer stem cell model, improved anti-cancer therapies may be identified based on the ability to kill these cancer stem cells rather than the bulk population of non-tumorigenic cancer cells. Alternatively, if cells with tumorigenic potential are common, it will not be possible to treat cancer more effectively or to understand cancer biology by focusing on small minority subpopulations.[9]

Studies have shown that ionizing radiation can kill most glioblastoma (a type of brain cancer) cells except those expressing a specific cell-surface marker, CD133. As it turns out, CD133 cells are resistant because they are more efficient at inducing the repair of damaged DNA.[10] Any stemlike glioblastoma cells would also be resistant to radiation.

Excited about the cancer stem cell idea, its advocates began lobbying the oncology community to refocus the NIH budget to support their research. Dr. Franklin was invited to give testimony at a Senate oversight subcommittee of the National Cancer Institute where cancer stem cells were discussed.

■ ■ ■

Scene: Genetic oncologist Dr. Arthur Cosgrove is trying to convince Senator Brad Furst, a senior member and chairman of the oversight subcommittee for the National Cancer Institute, to add a rider to the new NIH budget that dedicates $500 million for research on cancer stem cells, which Cosgrove believes will revolutionize treatments for cancer. The Senate hearing where Dr. Cosgrove has been called to testify is focused on the new NIH budget. The second person invited to testify is Dr. Franklin, for her expertise as a bioethicist who has written about the allocation of resources to health and medical research. It is well known that she turned her career around to find a cure for her father's paralysis, which brought her to the attention of Senator Furst.

FURST: Dr. Cosgrove, speaking on behalf of the committee, we are pleased that you have accepted our invitation to help us understand new developments in oncology that may need to be reflected in the new budget. We will enter your written comments into the record. Please raise your right hand and swear that everything you say before the committee is the truth as you understand it.

COSGROVE: I so swear, Senator.

FURST: You may begin.

COSGROVE: Senator, thank you for the opportunity to appear before your subcommittee. We in the field of oncology use the word "cancer" to describe a group of diseases that are characterized by uncontrolled cell growth. Some of these abnormal cells invade adjacent tissue or other parts of the body and cause secondary tumors—a process we call metastasis. The standard therapies we use to treat cancer include radiation, chemotherapy, surgery, and hormonal treatments. Often we do not rid the body of all the abnormal cells. Some will return in a few years. These techniques can also harm normal cells.

In the past few years scientists have discovered cancer stem cells. These are cells found in or near tumors that can self-renew and can also differentiate into the variety of other cells found in the tumor growth.

A very important new hypothesis has been adopted by some scientists: that the experimental and clinical behaviors of metastatic cancer cells resemble the classical properties of stem cells. This cancer originates from a small population of stemlike tumor initiating cells. If this hypothesis is correct, then cancer may arise from a mutation in a stem cell. Cancer stem cells are likely to be the origin of cancer metastasis.[11]

To stop the progression of cancer, we have to kill off these cancer stem cells. This is the most exciting new development in oncology in decades, and we need a special appropriation to move this research along without delay.

FURST: Dr. Cosgrove, I can hear the passion and the urgency in your voice. I have been on the NCI oversight committee for thirty years and have heard many passionate voices in support of the latest potential cure for cancer. There was a period when virology was believed to be the answer. For a while after the virology idea waned, there were proposals for the "magic bullet" approach to killing cancer cells without harming normal cells. This was usually a targeted molecule or virus with a toxin that would destroy tumor cells. More recently, angiogenesis was going to reveal how we can eradicate tumors and stop their spread by killing off their blood supply.

I know we need a multifaceted approach because we are not dealing with one disease. And we don't know where we're going to find the answer. So you can see that, having been in this seat for so long, I have developed a cautious skepticism about cancer therapies, until I see them work. Do you have anything more than promise?

COSGROVE: Senator, what makes this new approach so exciting is the new science behind it—not the trial and error that has characterized so many other approaches to cancer therapy. The idea of a cancer stem cell helps us understand why we have made false starts. Oncologists have been perplexed about why chemotherapy and radiation destroy cancer cells but the cancer can still grow back. These treatments did not eradicate the progenitor cells, which we now know are the cancer stem cells.

FURST: Okay, Dr. Cosgrove. You make it sound like the idea of cancer stem cells has been proven. Some of our NCI scientists testified yesterday that this is still a hypothesis. You are asking Congress to invest a substantial amount of public funding, based on one hypothesis.

COSGROVE: With all due respect to your fine NCI scientists, I am here to tell you that cancer stem cells are more than just a hypothesis. Scientists have discovered stemlike cells in solid tumors of the breast, colon, brain, pancreas, and prostate. These have to be tested to establish definitively their role as cancer stem cells.

FURST: This leaves me with two questions. How do you intend to prove that these stemlike cells are the real thing? And what can you do with this knowledge to help people with cancer?

COSGROVE: We have to be able to separate the cancer stem cells from the ordinary, more plentiful cells. We then treat mice with both sets of cells. If the mice develop tumors exclusively from the cancer stem cells, we know we are on the right track. In our view, the cancer stem cells are resistant to traditional therapies, or can repair themselves after those are used.

FURST: And how will this knowledge help clinical oncology treat cancer?

COSGROVE: Once we distinguish and classify these cancer cells, we will develop strategies that will interfere with the molecular pathways that increase drug resistance to the cells and target cell proteins that will sensitize the cells to radiation. We will determine how to interfere with the cells' proliferative capabilities and block their ability to make tumors. To quote one group of researchers, "There are several ways of combatting CSC activity including inducing apoptosis [cell death], inhibiting stem cell self-renewal to either stop their division or to promote their differentiation, or targeting the CSC niche that supports them."[12]

Senator, I can leave you with this important finding. Cancers are perpetuated by a small population of tumor-initiating cells that exhibit numerous stem cell-like properties. We need NCI to provide adequate support for scientists to bring this discovery from the laboratory to clinical practice. Thank you.

FURST: Thank you, Dr. Cosgrove, for your passionate and informative testimony. We will enter your written comments into the record.

The next panel member is Rebecca Franklin, who is a bioethicist and stem cell researcher.

Dr. Franklin, we are very pleased you could attend this hearing. Few people doing bench research in stem cells also have a distinguished career in bioethics. Please raise your right hand and swear that everything you say before the committee is the truth as you understand it.

FRANKLIN: I do so swear.

FURST: Welcome to our subcommittee, Dr. Franklin. You have a distinguished career that includes receiving your medical degree and your doctorate in medical genetics and a master's degree in bioethics, which included studies in the history of medicine and medical research. You are also a distinguished researcher turning your attention to stem cells with a mission to create a cure for your father's paralysis. I can think of no person with a stronger reason to support medical science and with a better understanding of its history. Our subcommittee is part of the oversight function of the National Cancer Institute. Stem cells have become an important area of research, and the NIH is being asked to reorganize its budget to take account of the promise that they offer. But we are being advised to allocate $500 million to one piece of the stem cell promise—cancer research. If we make that allocation, then we are betting on one path of discovery, and others will be set aside for a future time. Can you offer the subcommittee your view of the stem cell promise and the fair allocation of government resources, particularly during this economic downturn?

FRANKLIN: Senator, thank you for the kind introduction. It is an honor to appear before the subcommittee. It has always been my view that science must be allowed to pursue every ethical path that is open, because we can never predict which will yield a major discovery. Stem cell research has opened up many new paths of inquiry both for basic science and for translational medicine.

There are many exciting possibilities in the medical applications of stem cells. Some are already in Phase I clinical trials. It has been reported recently that a small number of stroke victims who were treated with stem cell injections have exhibited improved movement, and one victim was able to speak again.

In terms of the greatest current potential for treating disease with stem cells, I would have to say: stroke, macular degeneration, Parkinson's disease, spinal cord injuries, and multiple sclerosis rate highest on my list.[13]

FURST: So where does cancer research fit into this set of prospects?

FRANKLIN: The National Cancer Institute has the largest budget of the NIH institutes. Basic research in cancer has evolved over the forty years since the "War Against Cancer" was launched in 1971—through cancer virology, environmental oncogenesis, angiogenesis, and oncogenetics. There have been important translational applications of these research paths, most recently, the use of gene analysis to identify classes of tumors that respond to specific drug treatments. There are also new alternative theories to the standard model of cancer, namely, the somatic mutation theory that cancer arises from the mutation of a single cell.

FURST: Are there other alternative theories?

FRANKLIN: The tissue organization field theory argues that cancer arises from the interaction of cells and tissue. Proponents have evidence that cancer is a field phenomenon involving a disruption of tissue-to-tissue communication. So they believe it must be studied at the level of tissues and organs rather than cell DNA.[14] We clearly need new approaches. One of the nation's leading cancer biologists recently wrote that we lack the conceptual paradigms and computational strategies for dealing with the complexity of cancer.[15]

FURST: What is your assessment of the cancer stem cell hypothesis?

FRANKLIN: The theory about cancer stem cells is at a very early stage and needs considerable validation before it can be widely accepted. It has some plausibility, but many scientists are skeptical of it because the markers for cancer stem cells have not been fully validated.[16] So in my view, there should be some funding for research to establish proof of concept. But I cannot see a rationale for more funding than for other areas of stem cell research that have a higher probability of succeeding at this time. Some are already in clinical trials. Once we see a few successes in stem cell technology, the public will realize its medical benefits and there will be a firestorm of public support.

FURST: Dr. Franklin, you have been an advocate of "stem cell pluralism." What do you mean by that?

FRANKLIN: Senator Furst, I believe in letting a thousand flowers bloom in science rather than putting all our resources into one area. We have lost many good ideas in cancer research by neglecting promising approaches that were not mainstream.

FURST: Isn't the stem cell hypothesis out of the mainstream?

FRANKLIN: Indeed it is. And in my view, research on it should be funded, but not at a level that overshadows other promising ideas in their early stages of development.

FURST: Thank you for your candid remarks. We shall include your full written testimony in the record of this hearing.

DIALOGUE 12

REPROGRAMMING CELLS

The reversal of organismic development has always been, like the movie *Back to the Future*, a science fiction fantasy. Can an adult be restored to youth? Can we reverse aging?

Aging, like time, seems to follow an entropic principle. It moves in one direction and cannot be reversed. But this might not be the case at the level of a cell, rather than an entire organism. Each cell in an organism can ultimately be traced back to a pluripotent stem cell, usually a multipotent cell, from which it was eventually differentiated. Can the adult differentiated cell be reprogrammed to the state of its forbear stem cell or even further back, to become a pluripotent embryonic stem cell-like cell?

The process of reprogramming an adult cell to its pluripotent stem cell origins, sometimes referred to as "dedifferentiation," creates what are called induced pluripotent stem cells (iPSCs). The first evidence of this type of reprogramming from mouse cells was provided by Shinya Yamanaka and his colleagues at Kyoto University in Japan in 2006.[1] A year later, in November 2007, Shinya Yamanaka and James Thomson, working independently, created iPSCs from adult human cells. Yamanaka and embryologist John Gurdon were awarded the 2012 Nobel Prize in Physiology or Medicine "for the discovery that mature cells can be reprogrammed to become pluripotent."

The induced pluripotent stem cells were seen to possess at least two advantages over embryonic stem cells. First, because the iPSCs could be derived from a patient's adult cells, they could be used in treatment without any adverse immunogenic responses.[2] Second, because pluripotent stem cells are derived from somatic cells, such as skin cells, there is no need to destroy human embryos.

Initially, iPSCs were produced by transferring four genes (called transcription factors) by viruses (transfection) into an adult cell. After several weeks a small number of adult transfected cells became morphologically and biochemically like the pluripotent stem cells derived from embryos.

By 2009 scientists had learned to reprogram adult cells without making any genetic alterations. They accomplished this by repeated treatment of the cells with certain proteins.[3] Because the American government's restrictions on the use of embryonic stem cells have been so severe, many scientists seeking federal research grants redirected their studies to the use of iPSCs. According to Alicea and Cibelli, there remains considerable mystery about how the reprogramming works.

> The mechanisms underlying such dramatic transformation are poorly understood. In many ways, the science of direct and indirect cellular reprogramming is like poking a wild animal, e.g., a bear, with a stick. Like the bear, a cell has a complex internal state that responds to external perturbations in a highly non-linear fashion. In fact, predicting the response to a poked bear is far easier than predicting the response of a cell undergoing reprogramming, largely because the bear's behavior is better characterized.[4]

Dr. Franklin discusses the pros and cons of moving away from embryonic stem cells to iPSCs with a leading stem cell scientist.

■ ■ ■

Scene: Dr. Frederick Jones is a stem cell biologist who has invested considerable time reprogramming ordinary cells to make them into pluripotent cells. Dr. Franklin questions him about the reversibility of cellular life—playing the role of skeptic about the prospects of reprogramming cells to produce therapeutic stem cells.

FRANKLIN: U.S. policies limiting federal support for research on embryonic stem cells have had an unanticipated effect. Scientists who felt constrained by President George Bush's policy on stem cells began investigating alternative approaches to acquiring pluripotent stem cells that do not involve destroying a human embryo.

Yamanaka in Japan, who, like President Bush, had ethical concerns about destroying human embryos,[5] turned mouse skin cells into pluripotent stem cells in 2006. The scientific community was abuzz. How did that play out among stem cell researchers?

JONES: It was a funny thing. Some scientists were very excited and immediately jumped on the induced pluripotent stem cell bandwagon, especially those seeking federal funds, because the available cell lines were limited. Scientists who were in the private sector or part of privately funded research institutions, unconstrained by federal stem cell policy, were much more skeptical. Even if reprogramming were able to create embryonic-like stem cells, they raised many questions about whether these would be as effective as embryonic stem cells. They questioned whether iPSCs and blastocyst-derived ESCs were molecularly and functionally equivalent.

FRANKLIN: As I understand it, the next few years brought mixed responses to the question of equivalence and comparable plasticity, with similarities and differences between the two types of cells reported. For example, neural cells derived from iPSCs were more likely than embryonic stem cells to form tumors after transplantation into the brains of immune-compromised mice.[6] Also, the human iPSC-derived blood progenitor cells seem to undergo premature senescence, a foreboding sign for any possible therapeutic use.

JONES: It is certainly true that the first generation of induced pluripotent stem cells had important genetic and epigenetic differences from their embryonic counterparts, one of which is that iPSCs are more likely to be tumorigenic. But we have to understand that the initial discovery of reprogramming adult somatic cells is a proof of concept. We have learned that the plasticity of somatic cells is much greater than was originally understood. In addition, the discovery offers a possible solution to the moral objections to destroying early human embryos. There are still many possibilities for creating iPSCs from somatic cells that are more like their embryonic counterparts.

FRANKLIN: Ironically, even if iPSCs become more like ESCs, we are not out of the woods. Embryonic stem cells themselves show a high degree of oncogene expression and tumorigenesis. So is it a matter of degree?

JONES: We stem cell scientists are still trying to understand how the oncogenes are turned on and off in embryonic stem cells and how they are affected by reprogramming. But you are correct: thus far, some scientists have observed that iPSCs seem to create more tumors than hESCs.[7]

FRANKLIN: From my embryology studies, I can fully appreciate the functioning of cells in an organism compared to their development in culture. A number of studies have shown that culture-adapted human embryonic stem cells can form more aggressive tumors.[8] The reprogramming process often introduces not only genetic abnormalities but also epigenetic alternations, which are expressed in the tumorigenicity of cells.[9]

JONES: There have been many breakthroughs in biology that seemed unrealistic when the research began. Reprogramming somatic cells is very new, and we're not going to get it right on the first try. I do not agree that biological systems are inherently irreversible. We simply have to break the code, as Crick and Watson did when they discovered the structure of the DNA molecule over fifty years ago. For a long time organ transplants were impractical because we couldn't decode the immune system's rejection of foreign bodies. Doing so was a kind of reprogramming. Today we have drugs that suppress the body's rejection of foreign tissue.

FRANKLIN: But unlike in the examples you cite, the more we learn about iPSCs, the more roadblocks to possible clinical applications we discover. The reprogrammed cells not only are more tumorigenic than their counterpart embryonic stem cells but also exhibit more mutations and many changes in their epigenetic profile. Andras Nagy, at Mount Sinai Hospital in Toronto, found that induced pluripotent reprogrammed adult cells have more genetic errors—about three times more than embryonic stem cells. This makes them less desirable for regenerative medicine.[10] Another study, published in *Nature*, found that iPSCs contain ten times the expected number of mutations. And adult cells do not convert well to an embryonic-like state. Patterns of one of the important epigenetic changes in DNA, the binding of methyl groups to DNA molecules, near the tips and centers of chromosomes of the reprogrammed iPSCs resembled methylation patterns in the adult tissues from which the cells had been derived.[11] In other words, the reprogrammed cells have the epigenetic memory of the adult cells, and this could limit the types of tissue the iPSCs are capable of forming.[12]

JONES: We do not yet know whether these roadblocks to regenerative medicine are insurmountable. We might be able to reset the methylation patterns so we remove the memory. Or if the iPSCs are differentiated, the epigenetic marks might disappear—sort of an induced memory loss. These effects may also be an artifact of the culturing techniques used to derive and maintain the cells. As those techniques improve, so may the quality of the cells. Even as we recognize the challenges of applying iPSCs and embryonic stem cells, the knowledge we are gaining is quite remarkable.[13]

FRANKLIN: From the perspective of an organicist-embryologist, I can certainly appreciate the fact that cells cultured and maintained outside of a normal organism can acquire factors that may not be compatible with the environment of the human body, where the signals are so much more complex, once they are transplanted.[14] Ironically, there are roadblocks to regenerative medicine not only with iPSCs but also with embryonic stem cells when they are grown in culture media. Such cells acquire an abnormal number of chromosomes, or become aneuploidy.[15] Overall, studies have shown that the reprogramming and subsequent expression of iPSCs in culture can result in diverse abnormalities compared to the cells from which they originated. The medical importance of these abnormalities will have to be studied in great depth before these cells can be used in restorative medicine. Currently, it seems to me that induced pluripotency is not a substitute for the use of embryonic stem cells, and we do not know whether epigenetic differences will be consequential in therapeutic uses of the iPSCs.[16]

JONES: Well, I don't share your skepticism. We have just begun turning somatic cells into pluripotent stem cells. The refinements will address some of your concerns. How iPSCs ultimately compare with embryonic stem cells will eventually be resolved by experimental science.

DIALOGUE 13

MY PERSONALIZED DISEASE CELLS

The discovery of induced pluripotent stem cells (iPSCs) has opened up a new path for the study of diseases and effective therapies to treat them. Certain diseases result from abnormal cells. By reprogramming abnormal cells from an adult afflicted with a disease using iPSC methods, scientists can re-create the precursor embryonic-like stem cells. These can then be grown, cultured, and differentiated to produce a continuous supply of diseased cells. Studying them will help scientists understand the development of the cell from its embryonic-like state to its mature form found in a particular tissue. For example, culturing neurons from adult animals has been historically difficult.[1] Stem cell technology may improve the success of cell cultures for many types of abnormal cells, as noted by Park et al.[2]

Tissue cultures of immortal cell strains from diseased patients are an invaluable resource for medical research but are largely limited to tumor cells or transformed derivatives of native tissues. Induced pluripotent stem cells are generated from patients with a variety of genetic diseases arising from either a single gene (Mendelian) or a complex genetic inheritance. The diseases for which iPSC lines are of value for research include adenosine deaminase deficiency-related severe combined immunodeficiency (ADA-SCID), Shwachman-Bodian-Diamond syndrome (SBDS), Gaucher disease (GD) type III, Duchenne (DMD) and Becker muscular dystrophy (BMD), Parkinson's disease (PD), Huntington's disease (HD), juvenile-onset, type 1 diabetes mellitus (JDM), Down syndrome (DS)/ trisomy 21, and the carrier state of Lesch-Nyhan syndrome. Such disease-specific stem cells offer an unprecedented opportunity to recapitulate both normal and pathologic human tissue formation in vitro, thereby enabling disease investigation and drug development.[3]

If a drug can induce the cell to function normally, then it can be tested on the culture of diseased cells before any testing is done on an individual. Scientists call these "diseases in the dish." For example, a number of childhood diseases are expressed by abnormalities in brain development. Scientists have thus far not been able to obtain brain cells such as neurons, astrocytes, or microglia in culture. Using iPSC technology, skin cells can be reprogrammed to pluripotent stem cells, which can be differentiated to form brain cells. These cells will carry the genetic deficiency of the patient. Scientists can then explore drugs that could repair or minimize the defect in vitro, before they are tried on the patient. To study neuron demylenation or cell death, scientists can culture the defective cells and model the disease in the dish rather than in an animal.

Another example of "disease in the dish" is muscular dystrophy (MD), in which muscle cells and other cells in the body cannot bind to laminin. The laminin family of glycoproteins is a central part of the structural scaffolding in almost every tissue of an organism. It is vital for the maintenance and survival of tissues. Defective laminins can cause deformation of muscles, skin-blistering disease, and defects of the kidney. Binding to laminin allows the muscle cell to connect to the extracellular matrix. Without binding, the muscle cell cannot function properly.

The laminin binding assay shows which cells can or cannot bind and to what degree. This test will allow scientists to determine whether any existing drugs will increase the laminin binding in cells. The results are not in yet, but this is the current approach in the "disease in the dish" strategy. Diseases of neurodegeneration, such as Alzheimer's and Parkinson's, involve cellular death and dysfunction. They can now be modeled in a dish with reproducibility, so scientists no longer must rely on rat neurons to do this research.[4] For example, in his essay titled "Diseases in a Dish," historian Stephen S. Hall describes how an ALS patient donated a skin sample for research at Columbia University, where a medical geneticist planned to apply iPSC methods to turn the somatic cells into motor neurons.

If successful, the disease-in-a-dish concept could speed up research on many different diseases and lead to faster, more efficient screening of potential drug therapies, because scientists would be able to test potential drugs in custom-made cultures for both therapeutic efficacy and toxicity. Induced stem cells are currently being used experimentally to model dozens of illnesses besides ALS, including sickle cell anemia, many other blood disorders, and Parkinson's disease.[5]

In the next dialogue, scientists discuss this approach for studying congenital cardiac problems.

■ ■ ■

Scene: Dr. Leonard Phillips is a pediatric cardiologist who is treating a young child with an unusual arrhythmia. Dan Henderson is a cell biologist whose research centers on cardiac electrolyte abnormalities, one of the causes of arrhythmia. Henderson has embraced using stem cells as a revolutionary approach to evaluating drug therapies for electrolyte abnormalities that have a genetic etiology. Phillips and Henderson meet after a day-long conference on new frontiers in cardiology. Dr. Franklin and Dr. Phillips were in the same medical school class, and Franklin attended Henderson's talk at the conference. They all meet at the hotel bar.

PHILLIPS: Dan, I heard your talk on developing a stem cell approach for cardiac disease. I want to discuss a possible collaboration. I have a patient who could possibly benefit from your research.

HENDERSON: Tell me about your patient.

PHILLIPS: Troy is a four-year-old with a congenital arrhythmia. We have traced it to an abnormality in his ion [sodium, potassium] channels. Thus far we have provided him with an internal cardiac defibrillator (ICD), and he is also on drugs that are not fully effective.

HENDERSON: Have you studied the underlying cause?

PHILLIPS: We screened his DNA for a dozen candidate genes that are associated with electrolyte channels and discovered one polymorphism and a mutation that we believe is the cause of his problem. After hearing your talk, I realized that you might be able to use Troy's skin cells to create a pluripotent stem cell that could be induced to make cells that are surrogates for his heart cells.

HENDERSON: What we have been able to do in a similar case is obtain skin fibroblast cells from a patient and the patient's family. Then, using induced pluripotency, we were able to return these cells to an embryonic-like stage and induce them to differentiate into cardiac cells. Now with cardiac cells of parent and child, we can study the single cell's electrophysiology. The resulting iPSCs hold the developmental potential of human embryonic stem cells without the embryo. This would give us information about the dysfunction in Troy's cells by using his parents' cells as controls.

PHILLIPS: What are the constraints?

HENDERSON: We begin by studying certain inherited arrhythmogenic diseases caused by simple mutations or polymorphisms in which there is a clear relationship between genotype and phenotype. We have created cardiomycytes [cells that form cardiac muscles] from iPSCs. These contain the same disease-causing mutations that are present in the heart tissue that produce the clinical symptoms.[6]

PHILLIPS: What do you do with these cells?

HENDERSON: We get their full electrophysiological characterization, such as specific ionic currents and action potential profiles.

PHILLIPS: Okay, suppose you now can understand how the cells respond to different electrolytic changes. What's the next step?

HENDERSON: We then can determine which available or new drugs change the electrophysiology of the cell to a more normal function.

FRANKLIN: Are you saying that you create a model from the patient's own cardiac-like cells and use those cells to test new drugs?

HENDERSON: That's essentially it.

FRANKLIN: How can you know whether the cells studied in vitro will function the same when they are in the patient?

HENDERSON: That's always the challenge in cell biology. We determine from cells in vitro which antibiotics will be effective. But we have to test in the animal and in the person to see if the model system will work in the organism.

PHILLIPS: If your iPSCs help in selecting a drug that allows us to stabilize the ion channels, then we can remove the patient's implanted defibrillator.

HENDERSON: That's down the road. We view this research as proof of concept. If the method works, it can be generalized to other cells in other parts of the body. We are interested in creating personalized cells for any tissue of the body that can be tested in vitro for their response to drugs—if drugs are available—to treat disease even if the etiology of the disease is genetic. The common myth is that someone who has a genetically caused disease cannot be treated.

FRANKLIN: It seems that you don't even have to use animal models.

HENDERSON: We do use animals to validate the iPS methodology, but the beauty of this system is that we test the drugs on human cells—and not only that, on cells that mimic the target cells of the patient, their personalized disease cells. We could never get this close to the patient in an animal model.

Our mission is to get therapies to patients—through small molecule therapies by understanding disease processes better; through human toxicity studies that are more predictive and save massive amounts of time and money compared to animal studies; and through cellular therapies that we can move more quickly into clinical trials.[7]

FRANKLIN: Have you been approached by animal rights organizations? Wouldn't it be more humane to reduce as much as possible the number of animals sacrificed for research?

HENDERSON: Animal rights groups like PETA have been promoting the use of human cell culture as a substitute for animal studies for decades. One of their main arguments is that no matter how many animal studies you do, you cannot get a drug approved through the FDA unless you test it on humans. So why not begin there and save all the animal suffering?[8]

FRANKLIN: So what's your response?

HENDERSON: The ethical question is: Can we reduce the risks to humans by pretesting a drug on animals first? Some philosophers have argued that we should apply deontological ethics to humans—treat persons as ends in themselves—and utilitarian ethics for animals, weighing the benefits to humans when we decide how to treat animals.

FRANKLIN: How does that help advance the use of human cell culture?

HENDERSON: Well, if we can show that animal studies will save human lives and help us reduce the adverse effects of drugs before testing on humans, we can justify using animals. But if no such justification is possible, then I would say it is unethical to sacrifice animals for humans.

PHILLIPS: Of course, there is still a big gap between what works in vitro and what works in the patient. But we have entered a new realm of working with patients' tissue in vitro.

HENDERSON: There are a number of groups working on this methodology of personalized IPS cells. A group at Harvard started with skin cells from patients with cystic fibrosis [a genetic lung disease leading to respiratory failure] and created induced pluripotent stem cells. They used those cells to create the disease-specific tissues that line the airways, which is the most lethal aspect of CF.

FRANKLIN: How generalizable is their work?

HENDERSON: The same epithelial cells are involved in a number of common lung conditions, including asthma, lung cancer, and chronic bronchitis. So this research may result in the development of new insights into and treatments for those conditions as well.

FRANKLIN: It looks like the "disease in the dish" approach is beginning to gain some traction.

DIALOGUE 14

TO CLONE OR NOT TO CLONE: THAT IS THE QUESTION

The English word "clone" is derived from an ancient Greek homonym that refers to a process of creating a new plant from a "twig," the literal meaning of the ancient term. In modern biology, cloning describes a process of replicating or producing multiple copies of DNA molecules (molecular cloning), many copies of genes (genetic cloning), unicellular organisms (cell cloning), or whole organisms (organism cloning). DNA or gene cloning is accomplished by transferring a gene from an organism (using recombinant DNA techniques) into a self-replicating bacterium, which uses the machinery of its cell to make copies of the gene. A cell that divides by mitosis creates a cloned copy of itself. Artificial twinning of embryos also produces clones. Thus, to clone means to create a biological copy of DNA, a cell, or a living organism.

An organism can be cloned at its embryonic stage by artificial embryo splitting, or the fertilized egg can split naturally and spontaneously. The first animal cloning was accomplished by the embryologist Hans Spemann in 1901. He split a salamander embryo into two parts and observed its development into two complete organisms.[1] In 1952, Thomas J. King and Robert W. Briggs transplanted nuclei from frog blastula cells into enucleated eggs. Ten years later, John Gurdon showed that he could replace the nucleus of an egg cell of a frog with the nucleus of a mature intestinal cell. The modified egg developed into a swimming tadpole.[2]

The first sheep, named Dolly, was cloned at the Roslin Institute in Scotland in 1996. The process used in cloning her and the tadpole is called somatic cell nuclear transfer (SCNT). The animal's nuclear DNA (two sets of chromosomes) is extracted from a somatic cell and inserted into an enucleated egg (in which one set of chromosomes has been removed) from the same species. The egg containing the mature nuclear DNA (two sets of chromosomes) is treated by chemical activation or an electric current to stimulate cell division. When it develops into a blastocyst, it is implanted in the uterus of the host animal, where it is gestated. The newborn will have the identical nuclear DNA of the animal that provided it and is therefore considered a clone. This is called reproductive cloning. Therapeutic cloning also involves the insertion of a somatic cell nucleus into an enucleated donor egg. The somatic nuclear DNA is said to be cloned in the egg. The egg is activated for division until it reaches blastocyst stage; then the pluripotent cells are removed and differentiated for research or therapy.[3]

■ ■ ■

Scene: Dr. Howard Chadwick is director of the National Center for Stem Cells (fictional) within the National Institutes of Health. The purpose of the center is help integrate regenerative medicine into the work of all the institutes. Dr. Chadwick is between a rock and a hard place with regard to embryonic stem cells: he recognizes their potential in science as well as the political firestorm they have produced. He calls upon Dr. Franklin to help him sort out his dilemma and provide an ethical consultation.

CHADWICK: Rebecca, it is good to see you again. I recall the last time we worked together; it was on the president's ethics advisory board, addressing questions of human genetic engineering.

FRANKLIN: Yes, Howard. We were barely out of medical school then, and there was nothing in our education to prepare us for the task. We were reinventing ethics for modern genetics in the wake of the discovery of recombinant DNA and the possibilities of human genetic engineering. I suspect you have inherited the next generation of ethical issues.

CHADWICK: Right! The current issues are more palpable because they are not just theoretical. They are right before us. Let me summarize where we are. I have to start with Dolly the sheep. When Ian Wilmut's group at the Roslin Institute at the University of Edinburgh cloned an adult ewe, they used the method of somatic cell nuclear transfer (SCNT).

As you may recall, they took the nuclear DNA from cells taken from the udders of a deceased ewe and implanted it into the egg of another ewe. They actually acquired 277 udder cells before they could get one to perform.

FRANKLIN: I recall the *New York Times* headline: "Scientists Report First Cloning Ever of Adult Mammal." The report went on to say that such techniques could take a cell from an adult human and use it to make a genetically identical "clone." There was a bit of a media frenzy.

CHADWICK: There were two immediate responses to Dolly from the scientific community. The first was that this was an incredible feat, showing that DNA from an adult cell can be reprogrammed when it is transferred to an enucleated egg. It opened up new fields of scientific inquiry into the development and reprogramming of DNA. The second response was that the public fear of cloning humans could be used to block scientific progress. The entire program of nuclear DNA transfer could be stopped.

FRANKLIN: Efforts to ban human cloning have continued since 1978, when David Rorvik's sensationalized book *In His Image: The Cloning of Man* was published. Rorvik and his publisher initially claimed that his book was nonfiction and that he helped a businessman find scientists to create a clone of himself. After the publisher was sued, the company conceded the book was fiction. Rorvik, on the other hand, still insists it's true and never testified under oath. He did concede that three minor characters had been made up.[4] His book jump-started the discussion of human cloning.

CHADWICK: My staff has followed the United Nations Declaration on Human Cloning. The General Assembly adopted it by a vote of 84 in favor, 34 against, and 37 abstentions. So there is still a lot of uncertainty internationally about what a ban on cloning would mean. The scientific community should make a clear distinction between cloning human beings and using SCNT to create embryos for research. Strictly speaking, the same technology is used, but the outcome is different.

FRANKLIN: How do you mean?

CHADWICK: Well, for one thing, in SCNT for making embryonic stem cells, the oocytes with the transplanted nuclear DNA exist for a few days. They are never implanted in a uterus and therefore can never become a human clone.

FRANKLIN: Granted, it is not the intention of the research community to turn the egg with the adult DNA into a clone of the adult from whom the DNA was taken. But the ethical issues remain. If the government supports this research on SCNT, the efficiency of the process will be improved. We will not have to use hundreds of eggs to obtain a successful clone. Eventually, someone will probably try to use SCNT to make a human clone, unless it is prohibited and there are high sanctions against it.

CHADWICK: What the scientific community has done is to create a nomenclature that distinguishes two uses of the products of SCNT. One is called therapeutic cloning and the other reproductive cloning. We can support the former for medical research and oppose the latter as inherently unethical. We developed this distinction in the president's ethics committee on human genetic engineering, and it seems to have been accepted by the American public.

FRANKLIN: Some groups in society believe that an enucleated egg, which has been reconstituted with a complete set of chromosomes from a somatic cell and activated in what some have described as an ersatz fertilization or parthenogenesis, has moral worth. I am not sure that nomenclature will solve the ethical problem. It is reminiscent of the distinction that was made between somatic cell and germ line genetic engineering, which allowed scientists to parse out the moral problems.

CHADWICK: Precisely. We created terms that distinguished two types of human genetic modification. One refers to genetically altering living human beings for therapeutic purposes, such as curing type 1 diabetes by transplanting modified islet cells into the pancreas to correct the defective cells. And the other refers to genetic modification of the egg, sperm, or embryo. The vast majority of scientists were behind human somatic cell engineering and opposed to germ line engineering.

FRANKLIN: As I recall, that distinction was supported by the president's ethics committee and the National Academy of Sciences. The field of human genetic modification grew rapidly despite a generally accepted prohibition against genetically modifying the germ line.

CHADWICK: Why can't we create a similar distinction between therapeutic and reproductive cloning? We can achieve near universal consensus, with the exception of the transhumanists,[5] while eliminating any stigma against therapeutic cloning.

FRANKLIN: It is reminiscent also of the distinction that was advanced in the 1980s between gene therapy and enhancement genetic engineering. The former involves genetic interventions for curing disease and the latter genetic modifications for enhancing certain desired characteristics in a normal individual. The problem is that the distinction never held up. There was always a shifting boundary between therapy and enhancement.

CHADWICK: But the boundary between allowing an embryo to grow to term in a uterus and working with a donated IVF embryo for a few days to extract embryonic stem cells from the blastocyst is not fuzzy.

FRANKLIN: It may be a clear boundary for some outcomes, such as gestating a child, but fuzzy for others. Both reproductive and therapeutic cloning use living human embryos or embryolike cells. There is nothing fuzzy about the objective status of the human diploid egg.

CHADWICK: If the use to which an embryo is put does not justify a boundary, then the genetics and the physiology of the embryo's potential should. For therapeutic cloning we use an embryo that most likely cannot be gestated.[6] Scientists have discussed a process of disabling a gene that is necessary for producing the trophectoderm cells that protect the blastocyst; this removes the developmental potential of the reconstructed embryo.[7] It only has value for research and personalized medicine.

FRANKLIN: Whether or not you have a reproductively viable embryo, manipulating oocytes for research purposes, or what scientists called "parthenogenetically activating cells,"[8] namely, the development of an embryo without fertilization, is morally objectionable to some because human life is being created solely for human use.

CHADWICK: But surely we cannot consider a fertilized egg, or even eight cells from the progenitor cell, a human life. A living organism must be an integrated whole. These primitive cells have none of the qualities of a living organism.

FRANKLIN: Howard, we are in a multicultural world where the fertilized oocyte has a different moral and ontological status among different communities. In the Jewish tradition the embryo earns personhood after forty days in gestation; in Islam, this happens in the fourth month. In the Catholic tradition, personhood arises when the sperm meets the egg. Although they may differ on the timing, all afford the embryo some moral status that may limit its use to create stem cells.

CHADWICK: There is a long tradition of conflict between science and religion, stemming from the time of Galileo. Science cannot and should not navigate along the trajectory of religious doctrine. As a universal enterprise, science must act on its own. Early fertilized or reprogrammed oocytes are legitimate objects of scientific inquiry, and that inquiry should not be constrained by religious ideas.

FRANKLIN: Concerns about embryonic research transcend any particular belief structure. There is secular opposition to doing research on human embryos and engaging in therapeutic cloning. Remember James Thomson's remarks when asked about the ethical implications of his work: "If human embryonic stem cell research does not make you feel at least a little bit uncomfortable, you have not thought enough about it."[9] And Shinya Yamanaka said, "When I saw the embryo, I suddenly realized there was such a small difference between it and my daughters. I thought, we can't keep destroying embryos for our research. There must be another way."[10]

CHADWICK: Okay, I can understand that some people would be concerned about creating and then destroying an embryo. But there are countless numbers of embryos from IVF clinics that will be discarded. Isn't it better to find a humanitarian use for them? Why would a scientifically minded person be opposed to doing research on early embryos that would otherwise bedestroyed?

FRANKLIN: If a cloned embryo were created and—as researchers are attempting to do—could be grown for two months rather than the current seven to fourteen days, it then would be possible to harvest "embryonically derived germ cells."[11]

CHADWICK: Why would any scientist breach the fourteen-day limit when they can get as much pluripotency from the cells in the four-day-old embryo?

FRANKLIN: Now you are being naïve. The embryonic germ cells are believed to be nearly as versatile as embryonic stem cells, and they have one advantage:[12] they're less likely to induce tumors when transplanted into adult animals. So they are safer to use. Some secular scientists who do not view ESCs as revered human tissue see them as the first step toward working on more developed embryos, ultimately blurring the distinction between reproductive and therapeutic cloning.

CHADWICK: The problem with slippery slope arguments is that they can put a stop to all scientific progress. Anyway, how long an embryo is grown before being used for the derivation of therapeutic cells is less important than whether the embryo is implanted into the womb of a woman to grow into a baby (as in reproductive cloning). This is not a real slippery slope. The cloned embryo is either implanted into the uterus or it is not. There's nothing slippery about that.[13]

FRANKLIN: To address the slippery slope, voluntary restraints against extending the life of embryos are insufficient. Moreover, the current system in the United States involving one set of ethical rules for publicly funded stem cell research and another set for privately funded research is not justified except by political expediency.

CHADWICK: Embryonic stem cell applications are bound to improve. We wouldn't have to go further than an eight-to-sixteen-cell embryo.

FRANKLIN: But if they prove more difficult to use for therapies than producing iPSC-derived cells or tissues and organs from fetuses, there will be no shortage of biotechnologists, patient advocates, and supportive bioethicists who will encourage the public to adapt to such applications.[14] Scientists have already been extracting embryonic germ cells from a five-week-old embryo. These cells are also described as pluripotent and could have therapeutic value.[15] Meanwhile, many scientists believe therapeutic cloning must not be prohibited or we will end up in a place where we don't want to be.[16] I am only speaking now as a devil's advocate. Ironically, where we stand today, without therapeutic cloning I may not be able to achieve my goal of developing a patient-specific therapy for paralyzed individuals.

DIALOGUE 15

PATENTING HUMAN EMBRYONIC STEM CELLS IS IMMORAL AND ILLEGAL (IN EUROPE)

A German scientist, Professor Oliver Brüstle at the University of Bonn, filed for a patent on December 19, 1997, on neural precursor cells, the processes of their production from human embryonic stem cells, and their use for therapeutic purposes. The patent application claimed that the transplantation of neural cells into the nervous system is a promising method for the treatment of numerous neurological diseases, such as Parkinson's disease. The patent was granted in 1999. Greenpeace filed a lawsuit with the German Federal Patent Court (Bundespatentgericht) on the grounds that the precursor cells were obtained from human embryos, which were destroyed in the process. The Federal Patent Court ruled that the patent was invalid. The defendant appealed to the Federal Court of Justice (Bundesgerichtshot), which referred the case to the Court of Justice of the European Union (CJEU) for a preliminary ruling on three questions: What is meant by a human embryo?; Does the phrase "uses a human embryo for industrial or commercial purposes" cover the use for scientific research? and Can a patent be given to a product when its production necessitates the prior destruction of human embryos?

According to Article 27 of the European Union (EU) Agreement on Trade-Related Aspects of Intellectual Property Rights (signed on April 15, 1994), "patents shall be available for any inventions, whether products or processes, in all fields of technology, provided that they are new, involve an inventive step and are capable of industrial application." EU members may exclude inventions from patentability when they deem it necessary to protect public policy, public welfare, or morality, including to protect human, animal, or plant life or health or to avoid serious damage to the environment. In stark contrast, U.S. patent law, regulations, and practice have avoided making patent exclusions based on social and ethical concerns.

On March 10, 2011, the Advocate General of the CJEU offered a preliminary ruling in favor of Greenpeace and rejected patents on stem cells whose production involved destroying human embryos. The Advocate General argued that an invention cannot be patentable if the application of the technical process for which the patent is filed necessitates the prior destruction of human embryos for their use as base material, even if the description of that process does not contain any reference to the use of human embryos. The full Court of Justice of the European Union took up the case in Luxembourg and ruled on October 18, 2011. Their decision bars member states from awarding embryonic stem cell patents in Europe. The court found that "a process which involves removal of a stem cell from a human embryo at the blastocyst stage, entailing the destruction of that embryo, cannot be patented. The use of human embryos for therapeutic or diagnostic purposes, which are applied to the human embryo and are useful to it, is patentable, but their use for purposes of scientific research is not patentable."

The next dialogue addresses this decision. The characters are fictional and are not intended to resemble any real person. The statements attributed to them regarding the case *of Brüstle v. Greenpeace* are loosely based on the actual arguments in the case, as they were presented in court documents. However, they do not purport to represent the actual statements made by any party or individual.

■ ■ ■

Scene: Dr. Franklin moderates a panel discussion held at the Royal Society headquarters in London, about the international ramifications of the ruling by the Court of Justice of the European Union that patenting embryonic stem cells is unethical and therefore illegal. Any patents already awarded are to be withdrawn. Other members of the panel are Hans Weninger, a scientist at the University of Bonn whose colleague filed a patent on turning human embryonic stem cells into neural cells to treat neurological defects; Bettina Andrews, a lawyer and legal scholar specializing in the

European Union; and Jacques Penoir, a consultant to Greenpeace International and an activist bioethicist.

FRANKLIN: In the wake of the recent ruling by the Court of Justice of the European Union, we will be discussing the ethics and politics of patenting embryonic stem cells. [*Directing the question to Mr. Penoir*] Why did Greenpeace International file a lawsuit against a patent for a cell line that could be used to treat patients suffering from neurological damage?

PENOIR: Greenpeace is against the patenting or commercialization of human life. Greenpeace International has organized many public protests in Germany, including a 2004 rally at the Reichstag, and in other European states on the theme "Stoppt Patente auf Leben!" Human embryos represent human life.

FRANKLIN: Has Greenpeace opposed other patents in biotechnology?

PENOIR: Greenpeace International has consistently opposed patenting of human genes, transgenic animals, and plant germ plasm. Its campaigns against patenting seeds have slowed down the adoption of genetically modified plants in Europe.

FRANKLIN: How does Greenpeace reconcile its opposition to patents with the life-saving therapies that may thus be denied to people afflicted with degenerative diseases because, without patents, there is no incentive to develop such therapies?

PENOIR: We should remind ourselves that the Salk polio vaccine was not patented, yet it was available to countless numbers of people who were spared from the dread disease. A distinction can be made between patenting research and doing research. The EU decision does not ban research with human embryonic stem cells.

FRANKLIN: [*To Bettina Andrews*] How did the EU Court of Justice get this case? And why did it reach a decision that goes against the trend in many countries and a number of EU states that support research and commercial development of embryonic stem cells?

ANDREWS: Oliver Brüstle, a professor at the University of Bonn and Director of the Institute of Reconstructive Neurobiology, filed a patent application in 1997 on a method of making precursor neural cells from human embryonic stem cells, which are derived from blastocysts.

Brüstle's patent was issued by the European Patent Office in 1999. Greenpeace contested the patent in 2004 based on the clause "ordre public," which appears in the European Union's 1998 Directive on the Legal Protection of Biotechnological Inventions, to which all national patent laws of EU nations must defer. The European Directive on Biotechnology (Directive 98/44/EC) was intended to harmonize legislation on "patents on life" across Europe.

FRANKLIN: What is the power behind "ordre public" in EU law?

ANDREWS: The directive under "ordre public" states that inventions whose commercial use could lead to a breach of morality cannot be patented. The commercialization of the human embryo is cited as an example. Greenpeace built its legal challenge on that directive.

FRANKLIN: The original patent was taken out within the German patent system. How did the German courts address the Greenpeace challenge?

ANDREWS: In 2006, the German Federal Patent Court in Munich rendered a favorable decision to Greenpeace and declared patent claims on human embryonic stem cell lines illegal. Brüstle appealed his case to the German Federal Court of Justice. The German high court seemed to be leaning in Brüstle's direction but decided to refer the case to the Court of Justice of the European Union to clarify some ambiguous points of patent law on the definition and commercial use of human embryos and stem cells derived from them.

FRANKLIN: What specific questions did the German court want answered?

ANDREWS: With reference to Directive 98/44, the German court asked: Does the term "human embryo" include all stages of development of human life, beginning with the fertilization of the ovum, or must it meet other requirements, such as the attainment of a certain stage of development? Does the embryo also include somatic nuclear transplantation of a human ova, where a cell nucleus from a mature human cell has been transplanted into a human egg? Does it include unfertilized human ova whose division and further development has been stimulated by parthenogenesis? Are stem cells obtained from human embryos at the blastocyst stage included under the directive?

FRANKLIN: So that's how a German patent case got to the European Union Court of Justice. How did the court rule?

ANDREWS: On October 18, 2011, the CJEU ruled that "an invention is excluded from patentability where the implementation of the process requires either the prior destruction of human embryos or their prior use as base material."[1] Now it falls to the German Federal Court of Justice to make a ruling based on the recommendations of the CJEU.

FRANKLIN: I understand that the court ruled that any human ovum after fertilization must be considered a human embryo. What about if the ovum is not strictly fertilized by human sperm?

ANDREWS: The CJEU's decision included any nonfertilized human ovum in which has been transplanted the nucleus of a mature human cell and any nonfertilized human ovum whose further development is stimulated by parthenogenesis. This would incorporate the use of SCNT on human eggs to make embryonic stem cells.

FRANKLIN: [*To Penoir*] The goal of Greenpeace was to convince the EU to ban patenting of human embryonic stem cells based on its objection to the commodification of human embryos. But other countries like America, Japan, Israel, and Austria do offer patents for the cell lines. Aren't you concerned that the European scientists will flock to other countries to commercialize their discoveries, depleting scientific talent and wealth from Europe?

PENOIR: Ethics trumps economics. If the European Union law explicitly opposes the commodification of embryos, then Greenpeace was providing the EU with a valuable service by bringing it to the attention of the Court of Justice.

FRANKLIN: If there are no patents for human embryonic stem cells, then, according to the conventional wisdom, no company will make the investment to turn a discovery into a clinical product, and no translational research will take place in Europe. Therefore, European scientists will not be able to develop these cell lines to save people or improve the quality of their lives. Without patent protection, these goals will never be realized—at least not in Europe. Discovery alone will not bring practical therapies to people. Private companies must get involved. Banning patents denies people life-saving therapies.

PENOIR: This argument assumes that the end justifies the means. You can make the argument that all experiments on human subjects should be permitted, otherwise we may lose out on important clinical therapies. Or that we should commercialize living tissue and kidneys, if people would be willing to accept a high enough price for their body parts. Once we accept an ethical principle, we cannot undermine it by applying a utilitarian argument.

FRANKLIN: [*To Bettina Andrews*] Economic harmonization is one of the goals of globalization in general and one of the aspirations of the European Union. But ethics is culturally diverse within and across nation-states. If patents will be determined by ethical considerations, then they will likely vary among nation-states based on each of their moral considerations of the product.

ANDREWS: This is a major difference between the American and European approaches to patenting. The EU accepts moral arguments. The United States once did but no longer does. Ironically, for years neither reproductive nor therapeutic cloning research was eligible for federal funding in the United States. But federal patent protection was always available for those processes. A cloning patent was issued to the University of Missouri in April 2001 on methods for producing a cloned mammal and a cloned mammalian embryo. The U.S. federal government will not fund certain areas of research but will grant those areas of research funded by the private sector intellectual property protection. This has been called a "patent first, ask questions later" approach.

FRANKLIN: Was this always the U.S. policy?

ANDREWS: No. There was once a judicially created "moral utility" doctrine. This allowed both the U.S. Patent and Trademark Office and the courts to deny patents on morally controversial subject matter, under the assumption that such inventions were not useful. The U.S. Patent Act of 1952 closed the door on the moral utility doctrine. Currently, "no explicit basis exists for denying patent protection to otherwise patentable, morally controversial subject matter."[2]

FRANKLIN: What patents were denied based on the moral utility criteria?

ANDREWS: In the nineteenth and early twentieth centuries, applications for certain gambling devices were denied on moral grounds, and surprisingly, applications for nylon stockings with fake seams were denied because they were viewed as defrauding consumers.

FRANKLIN: Isn't it correct that that the Patent and Trademark Office denied a patent for a human-animal chimera that was submitted by a scientists and a biotech activist?

ANDREWS: Stuart Newman, a professor at the New York Medical College, and activist Jeremy Rifkin submitted a patent application on a laboratory-produced organism that was part human and part animal. It was putatively designed for medical research. The application remained active with the U.S. Patent and Trademark Office for seven years. In 2005 the office denied the patent because the subject matter under consideration (the human-animal chimera) was too closely related to a human and the patent office has a policy of not patenting a human being.[3]

FRANKLIN: Was its denial based on ethics—that it is unethical to patent a human being?

ANDREWS: In 1987, after the patent office issued patents on bacteria, yeast, and 436 animals, it said it would not patent a human, which would conflict with the 13th Amendment to the U.S. Constitution prohibiting slavery and would violate the privacy right of an individual, since it could preclude the individual from procreating.

FRANKLIN: Didn't the office grant patents on animals with human genes?

ANDREWS: Indeed it did. The issue is how much human DNA is in the hybrid.

FRANKLIN: How can it be legal to do research with human embryonic stem cells in Germany or other EU states, yet illegal to patent the cells?

ANDREWS: It is the inverse of the American question. How is it legal to patent hESCs but illegal to use federal funds to do research with them? Science, politics, and law are not always consistent. Dr. Brüstle called the EU court decision a disaster. He said, "It leaves European scientists with just basic research," while they watch as their discoveries are made into treatments around the world.[4]

FRANKLIN: Suppose there is a therapeutic use for the hESCs and it is patented in the United States. Can Europe accept the therapy?

ANDREWS: That is an open question. It is like asking, "Would you accept a therapy developed by Nazi scientists involved in unethical experiments?"

FRANKLIN: Personally, I believe that if it worked, most bioethicists would accept it based on utilitarian grounds. If an hESC therapy were not accepted by the EU nations, would any company anywhere be willing to back it in clinical trials?

ANDREWS: At least one U.S. economist said, "I don't see any company wanting to make money with a cell therapeutic that is going to ignore the European market."[5]

FRANKLIN: Will the U.S. Supreme Court decision that prohibits patents on naturally occurring DNA sequences have an impact on stem cell patents?

ANDREWS: That depends on whether the patentable materials are naturally occurring and the patent applicant has merely isolated them from nature, or the applicant has changed the cell structure. The unanimous decision by the Supreme Court in *Association for Molecular Pathology v. Myriad Genetics* held that "a naturally occurring DNA segment is a product of nature and not patent eligible merely because it has been isolated, but cDNA or Complementary DNA, an edited form of the gene, is patent eligible because it is not naturally occurring." The U.S. Patent and Trademark Office listed 2,137 patents with the term "stem cell" in the title by early 2014. This will keep a lot of patent lawyers engaged seeking to determine whether the patents will hold in Europe.

DIALOGUE 16

MY EMBRYO IS AUCTIONED ON THE INTERNET

The company eBay Inc. has revolutionized consumer trading, purchasing, and auctions online. Founded in 1995 by Tufts University graduate Pierre Omidyar, eBay is one of the successes of the dot.com explosion of the 1990s. The company describes its business model: "Buyers and sellers are brought together in a manner where sellers are permitted to list items for sale, buyers to bid on items of interest and all eBay users to browse through listed items in a fully automated way. The items are arranged by topics, where each type of auction has its own category."[1] Some have described eBay as the "perfect market," where transaction costs are minimal, sellers and buyers abound, and the company claims hundreds of millions of registered users of its services. In Adam Cohen's book *The Perfect Store: Inside eBay*, he states: "By connecting more than 30 million buyers and sellers around the world, eBay has permanently changed commerce. Things a buyer once would have spent days, weeks or a lifetime tracking down . . . are suddenly available at any hour of the day or night, from a personal computer in the buyer's home."[2]

After 13 years, company revenues reached $8 billion. eBay has incubated countless numbers of small entrepreneurs who use its penetration in a global market to establish an online store. Cohen writes that "as many as 100,000 people are already making their living selling on eBay . . . bringing people on the margins of the world economy into the economic mainstream."[3]

Among the items sold on eBay are previously undiscovered animal species, including an insect and a sea urchin. eBay conforms to regional laws and regulations regarding the sale or auction of products. Prohibited or restricted items include live animals and human parts and remains. Yet there is a rising demand for body parts. "Required for education, transplant and research, human tissues are in increasing demand. In the flourishing underground market, a single cadaver sold for its parts can fetch up to $200,000. Individual parts can make their way through several brokers before ending up in retail stores, fraternity houses, art exhibitions, the online auction eBay—or research labs."[4] In January 2007, the Louisiana Division of Archaeology was alerted by the National Park Service (NPS) of the sale of a suspect human skull identified in an eBay auction by a seller located in Lake Charles, Louisiana.[5] The sale was investigated by the Louisiana Department of Justice for violations of state or federal laws. Forensic anthropologists identified the skull as Native American. The seller was cooperative with investigators, and after he expressed his ignorance of the prohibitions against selling human remains, the charges against him were dropped.

Another case of an Internet auction of body parts had a different outcome. In December 2013 a suspect was arrested for allegedly stealing more than 60 jars of brain and other human tissues from a storage facility at the Indiana Medical History Museum. Court documents state that a California man claimed he had purchased six jars of the stolen brain matter for $600 on eBay.[6]

In the next dialogue, Dr. Franklin is in conversation with Janet Blackman, who has excess embryos from an in vitro fertilization procedure and wants to sell them on an Internet auction site. The views attributed to Janet Blackman are not those of any real person or company and do not reflect the official policies of any real company. Most companies, including eBay, comply with all applicable laws and regulations regarding the sale or auction of products, and their terms of use specifically prohibit the sale of human body parts. The Internet auction company in the dialogue, EggAuction.com, is fictitious.

■ ■ ■

Scene: Janet Blackman is a lawyer and feminist writer who specializes in embryo-related law and wrote a book titled *Embryo Rights*. She is married and gave birth to one child naturally. After several years of trying to conceive a second child without success, she underwent in vitro fertilization (IVF). She produced a dozen eggs but in two attempts could not get pregnant. Then she learned that her father had been diagnosed with amyotrophic lateral sclerosis (ALS). She had her son's DNA tested and found that he also had the ALS mutation. She herself was told she had incomplete penetrance of the mutation and would not have the disease. Both her father and her son would face years of medical care and neurological decline, and Janet knew that her family did not have the resources to provide the needed care. She realized that the embryos being stored in the IVF freezer were unique and might be worth a lot to a research institute or a pharmaceutical company. So she decided to auction them off on EggAuction.com. Dr. Franklin heard about the unusual case and contacted Ms. Blackman, who was being sued by a right-to-life group.

FRANKLIN: Janet, you have been publicly maligned and attacked by several "right-to-life" groups as well as the mainstream press for posting a notice on EggAuction.com that you were selling your embryos to the highest bidder. As I understand it, the embryos you posted for sale are in a freezer at an IVF clinic. Can you explain what led you to this decision?

BLACKMAN: I have a father and a child with amyotrophic lateral sclerosis. It is a great tragedy in my family's life. First I learned about my father; then I tested my son, who also has the mutation. I felt I had to do something to help my family pay for the care that they will require while also helping scientists understand this dreadful disease. I know that my genetics (my genome) is capable of shedding light on some of the secrets behind ALS.

FRANKLIN: So why don't you just donate the embryos to research?

BLACKMAN: The health costs of having two family members afflicted by the disease will be astronomical. Both my father and my son were disqualified from health insurance coverage because their genes for ALS were considered a preexisting condition. This was before the Genetic Information Non-Discrimination Act (GINA) and the Affordable Care Act were passed. So I thought I would put my relatively rare embryos up for sale. The highest bidder in this auction can then use my embryos to create embryonic stem cells with the ALS genes.

FRANKLIN: Don't you feel any moral ambiguity about reducing a living embryo to a commercial product? Doesn't it debase human life?

BLACKMAN: To me, this is a triple-win situation. I make money to pay medical bills for a terrible disease that has afflicted my family. Scientists can study the disease in the dish, and a company can use the result to develop a drug for ALS that makes it money. And society gets a treatment for the disease. [7]

FRANKLIN: Why did you choose EggAuction.com?

BLACKMAN: I can reach an international market and also get some publicity for the auction, since I didn't see any embryos for sale on the site.

FRANKLIN: Perhaps the reason you don't see any embryos for sale is that they do not accept human body parts, which is clearly stated by some auction houses.

BLACKMAN: My embryo is not a body part. Sperm banks sell sperm. As advertisements in college newspapers clearly show, researchers will pay a high premium for women's eggs. I am merely satisfying a market need for embryos, sperm-egg combinations. By the way, I did see a report that a healthy human kidney was put up for auction on eBay and it drew a worldwide bid of $5.75 billion.[8]

FRANKLIN: That report may be outdated, since eBay currently states that it does not auction human body parts.

BLACKMAN: Regardless, there is certainly an active Internet commerce in human eggs. My sale of embryos is just at the head of the curve.

FRANKLIN: The scientists who might purchase your embryos and eventually be awarded the intellectual property of your genome may undertake therapeutic cloning with your DNA or that of your son by somatic cell nuclear transfer (SCNT) to create your disease in a dish.[9] Does it disturb you that scientists will use your DNA to create human embryos from which to harvest embryonic stem cells, which involves destroying those embryos?

BLACKMAN: I have no moral concerns about scientists developing embryos for research, as long as they do not allow them to develop beyond about fourteen days, when the primitive streak appears—the elongated band of cells that forms along the axis of a developing fertilized egg early in gastrulation, a forerunner of the neural tube and nervous system. If scientists could have helped save my father or son from ALS by working on early embryos—what some call pre-embryos—I would have jumped at the chance. Perhaps it is my moral intuition, but I do not believe that whatever the moral status of ex-vivo pre-embryos is, it preempts their use to save lives.

FRANKLIN: It appears that the right-to-life stakeholders have tried to intimidate EggAuction.com with a lawsuit to prevent the sale of your embryos. They have also targeted you on their website. How has this pressure affected you and your family?

BLACKMAN: The attacks on me have been very difficult for my family, especially my son, who has already had to endure grief amid the illness of his grandfather. My son has been brought up with the value that doing things to help others is a virtue. He cannot understand why some of his peers are making fun of him at school about the media attention. I do not believe EggAuction.com has backed off—to their credit.

FRANKLIN: As you know, there is an existing market for human eggs. IVF clinics and individuals advertise for them at prices as high as $100,000 per egg for women with certain profiles. How much do you expect to get for your embryos?

BLACKMAN: My market is not IVF patients, so it is difficult to tell. No one contemplating having a child by IVF would want an egg from a woman who had a grandfather and son diagnosed with ALS. I expect that pharmaceutical companies, biotechnology companies, or private stem cell institutes will want my embryos. There is still time before the bidding is closed; that's when the final bid will be determined.

FRANKLIN: The right-to-life group that filed suit against you and eBay is called Safe Embryos. They are applying a legal theory that you are selling a child, which is against U.S. state and federal law. If you bequeathed the embryo to an institution or individual, there would be no problem. The lawsuit would be moot.

BLACKMAN: I am already in considerable debt. Why should I give away my personal property, my ALS embryo, to a wealthy drug company to be turned into intellectual property? If they pay dearly for the embryos, then they will have an incentive to earn back their investment by studying the development of ALS and perhaps finding therapies that will stop the mortality from it.

FRANKLIN: Are you aware of how the courts have ruled on ownership rights over human cells?

BLACKMAN: I am well aware of the story of Henrietta Lacks. Her cervical cancer cells were immortal and provided science with a very powerful cell line that was used to test the polio vaccine. But neither she nor her family shared in any of the wealth created by her cells. I am not about to let that happen.

FRANKLIN: John Moore's cells were taken from him during an operation. They proved very valuable for research and were patented. Mr. Moore filed suit to receive some of the financial rewards from the cell line that came from his body.

BLACKMAN: My situation is different than Moore's. The California Supreme Court treated his cells as abandoned, or at least outside of his property sphere, after they were removed by the surgeon. That is not the case with my embryos. They are under the custody of my husband and me, and under the contract we have the power to do as we wish with them.

FRANKLIN: The Supreme Court and lower courts have ruled that embryos are not persons. Therefore, the courts cannot rule that you are selling a child. We also have national laws that prohibit the sale of organs. You cannot put a kidney up for auction in the United States.

BLACKMAN: But the frozen embryos are not part of my body and cannot be considered organs. So I do not believe the plaintiffs will pursue organ laws on noncommercialization to defend their argument. And if they try, it will be dismissed in summary judgment.

FRANKLIN: The plaintiffs call you a baby killer on their website. Does that disturb you?

BLACKMAN: The plaintiff stakeholder groups make no distinction between an embryo and a baby. I haven't seen any attacks on me based on "organ commerce." So it strikes me that they want the courts to extend the law against selling children to prohibiting the sale of embryos. Sure, I am disturbed to hear them rant about a six-day-old embryo being a baby. They have created their own reality, and they build their law and ethics around it.

FRANKLIN: Some states have passed laws that prohibit the sale of live embryos. In 2010 the State of Florida passed Title XLVI Crimes,

Chapter 873, Sale of Anatomic Matter that prohibited the advertising or sale of human embryos.

BLACKMAN: But there is a caveat. It states, "No person shall knowingly advertise or offer to purchase or sell, or purchase, sell, or otherwise transfer, any human embryo for valuable consideration." But "valuable consideration" does not include the reasonable costs associated with the removal, storage, and transportation of a human embryo. So even in Florida, it's possible to receive money for the "donation" of an embryo.

FRANKLIN: Do you think your eBay auction will do better outside of the United States?

BLACKMAN: Perhaps. In 2006, the *Daily Mail* of London reported an embryo bank in the UK that sells its embryos for about 5,000 pounds each.[10] My embryos are uniquely important for medical research. So that is a different market. There is no doubt that as the value of embryos rises, there will be a global market for them, whether or not they are banned in specific countries. Remember, there is no law prohibiting the sale of human sperm or eggs. Fertilized eggs and embryos have not been banned from commerce either. Once the pharmaceutical giants recognize how valuable my fertilized eggs are, we shall see the EggAuction.com markets reacting.

DIALOGUE 17

HERE COMES THE EGG MAN: OOCYTES & EMBRYOS.ORG

In his essay titled "Regulating Markets for Human Eggs," David Resnik describes an unusual sale that became a model for the next dialogue. Resnik wrote: "In the fall of 1999, Ron Harris held a human egg auction on his website (www.ronsangels.com). The site featured three female models with accompanying photos and descriptions. Bids started at $15,000, with $1,000 increments."[1] Human eggs are sold or auctioned off in the United States without regulatory oversight. According to Rudhika Rao, "No federal law limits compensation for egg donors, and only a handful of state statutes address the issue directly."[2] Louisiana explicitly prohibits the sale of human eggs (oocytes), while Virginia authorizes their sale. Offerings for oocytes have been known to go as high as $100,000 for those from women of a select phenotype and educational background.[3]

There are widely divergent views about whether human eggs should be sold like any other consumer product. Some bioethicists make a distinction between eggs sold for research and eggs sold for in vitro fertilization (IVF). The April 2005 National Academy of Sciences *Guidelines for Human Embryonic Stem Cells Research* recommended that no payments, whether cash or in kind, should be made for eggs to be used in research.[4] In contrast, the International Society for Stem Cell Research, the American Society for Reproductive Medicine, and New York's Empire State Stem Cell Board have maintained that it is unfair to ask women to undergo the risks of egg extraction for no financial reward.[5]

The United Kingdom and France prohibit the sale of human eggs, but in the UK there can be reimbursement to the egg donor for time and inconvenience, which decommercializes the oocyte.

The new demand for human eggs in research, personalized medicine, and in vitro fertilization has created a livelihood for egg brokers. The broker cannot be duly compensated if the donors are not. Resnik notes, "Egg brokers are interested in compensating women for their oocytes, not just for their services. Why else would someone offer to pay [tens of thousands of dollars] for an egg from [a] tall, athletic, and intelligent woman?"[6]

In 2006 the Abraham Center of Life opened a commercial embryo business in Texas; however, the business plan failed and they closed within a year. Another company, California Conceptions, offers full-service infertility treatments that include embryos provided by paid donors. A *Newsweek* article quoted the company's description of its services: "These embryos are not the happy leftovers from another couple's quest to get pregnant, but are created for the purpose of providing them to patients, who pay for the entire procedure."[7]

The California Assembly (May 2, 2013) and the Senate (July 1, 2013) passed bill AB 926, which would have permitted the sale of eggs and embryos in the state. The bill was vetoed by Governor Jerry Brown on August 13, 2013.

■ ■ ■

Scene: Viktor Morovsky is a recent Russian immigrant who dreams of making a great deal of money in the American capitalist system, as a distributor for oocytes or embryos that people who have undergone IVF procedures no longer want. His company, Oocytes & Embryos.org, is preparing to go international and to be the largest single provider of embryo and oocyte sources for embryonic stem cells. He is in discussion with Arthur Worthington III, a venture capitalist who has invested lucratively in biotechnology and has made billions in BioVenture Associates. He brought Dr. Franklin along to help him understand the technical parts and the ethical implications of the business plan.

WORTHINGTON: Good morning, Viktor. Welcome to BioVenture Associates. I would like you to meet Dr. Rebecca Franklin, who is a consultant to our venture capital firm. I read your business prospectus and felt it was promising enough to hear more about, so that I could determine whether to bring it to the board of BVA. Can you give me the oral executive summary of your business plan?

MOROVSKY: The idea for this company was incubated in Soviet Russia. At that time, and to some extent in post-Soviet Russia, in vitro fertilization was introduced as a service to the government and military apparatchiks whose wives could not get pregnant. When I was a technician in the IVF clinic in Moscow, I was in charge of the cryogenic preservation of the unfertilized and fertilized eggs. For every egg used, there were a dozen frozen and kept in storage with no terminal date. Many of these eggs and embryos were discarded after a period of time. Then came stem cell research. All of a sudden these eggs and embryos became a valuable resource for regenerative medicine. Without eggs there are no embryos, and without embryos there is no research on embryonic stem cells!

WORTHINGTON: Were you able to make any money from this?

MOROVSKY: With Perestroika and the fall of communism, there was a free-for-all. I established a black market for these eggs when they were no longer needed. I made enough money to convince me that this was a lucrative business. There were no legal structures in postcommunist Russia to handle this business venture.[8]

WORTHINGTON: But there is a fragmented international market for selling oocytes and IVF embryos. Some European countries ban this practice as commercializing life. We just have to look at the decision of the Court of Justice of the European Union in *Brüstle v Greenpeace*, that embryonic stem cells cannot be patented.

MOROVSKY: Patenting the cells is one thing, but doing research with them and developing therapies is still very much actively pursued in Europe. And for this they need cell lines of embryonic stem cells. Just as there was a commercialization of HeLa cells for research in the 1950s, there is a growing demand for embryonic stem cells. I plan to be the major distributor.

FRANKLIN: How will you deal with the activist groups that are in opposition to paying egg donors? They argue that the commodification of eggs degrades personhood and dehumanizes human life. A number of countries prohibit or limit payment to egg donors for research, even though they do not limit payment of egg donors for IVF. Countries including South Korea, the United Kingdom, Canada, Australia, France, Germany, and Israel proscribe payment to egg donors for research on embryonic stem cells. The U.S. National Academy of Sciences recommends limiting payments to reimbursements for expenses. The state of Louisiana has an all-out ban on the sale of oocytes.

MOROVSKY: The main argument by critics, other than "commodification" of anything biological in commerce, is that poor women will have this incentive to risk hyperstimulating their ovaries. Currently, our business will only distribute excess fertilized or unfertilized eggs from IVF clinics. Thus, we will only be dealing with women who can afford IVF. Poor women who wish to sell eggs and do not use IVF will not be our providers.

WORTHINGTON: Will you be obtaining your eggs from international providers?

MOROVSKY: We are particularly fortunate for our business that some countries provide free IVF services to couples. Israel offers at state expense an unlimited number of cycles until the birth of two children. Germany also provides generous support for IVF. We expect that these countries will be a rich source of excess embryos.

FRANKLIN: Your business will create more demand for eggs. Poor women will find a way to sell eggs through the black market, whether directly to you or to others.

MOROVSKY: Dr. Franklin, the train has already left the station on payments to egg donors. Women donating their eggs for money is hardly a new thing. It has been happening in private egg donation banks throughout the United States and in Asia. Donating eggs is not an easy process. Women have to go through twelve to fourteen days of injections to stimulate hyperovulation, then a forty-five-minute surgery to obtain the eggs. Then the eggs are exposed to sperm, and after a number of procedures, embryonic stem cells can be extracted. So the sperm and egg donors should be compensated, although at different rates.

FRANKLIN: Selling eggs for IVF and selling eggs for research are dehumanizing in different ways. IVF eggs are used to build families. Therefore, the egg commerce is not predominantly for commercial purposes but to support new life. But research egg commerce is all about intellectual property, patents, and new drugs.

MOROVSKY: The Empire State Stem Cell Board decided in 2009 that women could be compensated for giving their oocytes or embryos to stem cell research. They concluded that there is no principled reason to distinguish between donation of oocytes for reproductive purposes and for research purposes when determining the ethics of reimbursement. The board said that donating oocytes to stem cell research confers a greater benefit to society than donating eggs for private reproductive use.

WORTHINGTON: The governor of California vetoed a bill that would permit the sale of eggs and embryos in the state. Does that indicate a trend, and how will that affect your business plan?

MOROVSKY: The governor acted quite unilaterally after there was a strong majority of the California Assembly and Senate that supported the bill. I believe the governor was out of step with the wishes of the people. There has been strong support from American bioethicists for allowing the sale of eggs and embryos for research and reproduction.[9]

WORTHINGTON: What about the international groups? Have you investigated their attitudes?

MOROVSKY: The International Society for Stem Cell Research does not condone payments for unfertilized or fertilized eggs, but accepts that some jurisdictions may allow them. Singapore approved compensation for egg donors in 2008, as long as it is not an inducement.

FRANKLIN: Will you be selling eggs for research exclusively or also for reproduction?

MOROVSKY: My business model is to sell and distribute fertilized eggs and embryos that have been in stored cryogenic facilities exclusively for research. The market for reproductive oocytes is very different. Women seeking to get pregnant by IVF will pay up to $100,000 for high-quality eggs; that generally exceeds the market rate for research embryos. Also, I do not have to carry the kind of insurance I would need if I were selling eggs for reproduction, where there are dangers to the women who donate their eggs and the eggs themselves may be defective. And the sale of eggs for reproduction gets into all sorts of racial issues that our business completely avoids. I am also counting on the fact that women with frozen embryos have to pay rent to keep them in the freezer. So if they do not plan to use them, I am doing them a service by compensating them for their frozen embryos.

WORTHINGTON: If the rules change about compensating women for the eggs or embryos they contribute to research, will you include that in your business plan?

MOROVSKY: I personally believe that donating eggs for stem cell research is a higher calling than donating for IVF, where the interests are usually personal and private or the interests of other couples who seek biological children.

WORTHINGTON: How so?

MOROVSKY: Egg or embryo donors who wish to advance stem cell research accept the risks of hyperstimulation to advance public welfare. They want their eggs or embryos to contribute to cures for diseases. If cultural attitudes change about compensating donors for research, we have a place in our business model to accommodate that.

FRANKLIN: What about the degradation of eggs or embryos kept in freezers for long periods of time? That could affect the quality of research.

MOROVSKY: The risks of eggs/embryos dedicated for research that are left in freezers and are no longer wanted for IVF are minimal. They will either serve or not serve the research function, but we have found that eggs remain robust for many years.

WORTHINGTON: What if researchers create embryonic stem cells from your fertilized eggs and those are used in human trials? Wouldn't you be liable if something goes wrong in the trial?

MOROVSKY: By that point there would have been so many hands on the fertilized egg that it would be impossible to tell whether the problem came from the original fertilized egg, the extraction of the pluripotent cell from the inner cell mass of the blastocyst, the process of differentiating cells from the embryonic stem cells, or the implantation procedure itself that could result in a failed effort or adverse outcome. We are neither the source of the fertilized egg nor the technicians who bioengineered them. We are simply the middlemen—the distributors.

WORTHINGTON: If egg/embryo donors are limited in what they can receive for contributing toward research, why wouldn't the limit be extended to frozen oocytes contributed for IVF use?

MOROVSKY: The purpose of the restriction on payment for egg or embryo donors for research is to eliminate monetary inducements that would influence women to undergo hormone treatments and egg extraction.

But the scientists who conduct embryonic stem cell research and the companies and universities that fund the research are all free to profit. So there are dual systems of ethics operating, one for IVF and another for research; one system of ethics for egg/embryo donors for reproduction and another for medical personnel and universities.

WORTHINGTON: Where do you expect to buy the oocytes/embryos?

MOROVSKY: They will be purchased in countries or states where it is not illegal to sell frozen oocytes/embryos that are no longer needed for IVF or oocytes obtained from ovariectomies. They will be shipped to an island country in the Caribbean to be processed for development of human embryonic stem cell lines. Those cells will be mailed all over the world to research laboratories, in the same way HeLa cells are distributed for research.

WORTHINGTON: What price will you give to women who sell you their excess eggs or embryos from the freezer?

MOROVSKY: The wonderful thing about the American free market system is that markets will emerge that set the prices based on supply and demand.

WORTHINGTON: What sales strategy will you use to persuade women to contribute their eggs/embryos?

MOROVSKY: The people who have the frozen oocytes/embryos know that if they don't sell them or give them away, they will die or be discarded. Our marketing firm will try to convince these women to donate their frozen eggs, whether fertilized or not, to research as an act of beneficence to support medical science. That is essentially what the courts have ruled when medical scientists extract cells from patients. The patient does not gain intellectual property value from the cells or tissue removed during an operation.

FRANKLIN: Let's be clear. You are not doing this for a higher public purpose. This is a for-profit enterprise.

MOROVSKY: While we are unabashedly a profit-making business, our service is consistent with a higher public purpose.

WORTHINGTON: Have you done a market analysis?

MOROVSKY: Markets have already begun, and we have analyzed them. A 1997 story in *The New York Times* reported that a U.S. fertility center was selling frozen embryos for $2,750, whereas embryo adoption centers were selling donated embryos for up to $10,000.[10] In 2000, the *Boston Globe* ran an ad that sought women ages 21–35 with at least one child to provide eggs for human embryonic stem cell research. The women were compensated between $500 and $4,000. I suspect researchers will be paying upwards of $5,000 for several eggs, and quite a bit more for eggs with a rare genotype. Under New York State's Stem Cell Science contracts, payments of up to $10,000 are allowable for oocytes designated for stem cell research. If I can sell 100,000 eggs annually with a profit of $1,000 per egg, that would amount to $100 million. As of 2002, a survey found that nearly 400,000 embryos were frozen in cryogenic storage facilities in the United States. While fresh embryos might be more desirable for embryonic stem cells, this source of embryos is underutilized commercially.

FRANKLIN: If you are referring to the RAND study, it stated that only 3 percent of the frozen embryos are designated for research. That's 12,000 embryos.[11]

MOROVSKY: My business plan includes an outreach to the couples who continue paying to keep their embryos in the freezer, which can cost upwards of $1,500 per year, when they are ready to sell the embryos for research.[12] A 2008 study of couples with frozen embryos who already had a baby found that about 3 in 10 were very unlikely or somewhat unlikely to use the embryos to have another baby.[13]

FRANKLIN: Some of the eggs/embryos have been frozen for an extended period of time. Do you expect to sell the eggs when they have been cryopreserved for ten years or more? They could be defective.

MOROVSKY: A group of scientists from Thailand reported that seventeen-to-eighteen-year-old frozen human embryos were thawed, cultured to the blastocyst stage, and induced to form embryonic stem cells. These frozen embryos retained their pluripotency, similar to fresh embryos.[14] Our potential source of embryos is much larger than we initially expected.

WORTHINGTON: Your business model looks promising, and you have addressed all the risk issues I can think of. We'll be back to you soon.

FRANKLIN: There may be some political naïveté that because society tolerates the commercialization of eggs/embryos for reproduction, it will also accept such commercialization for scientific research. That's the open question.

DIALOGUE 18

HUMAN-ANIMAL CHIMERAS AND HYBRIDS

The word "chimera" is derived from Greek mythology. The "Chimaira" was depicted as a fire-breathing female creature of ancient Lycia composed of parts of three animals: a lion, a goat, and a snake. The term has come to mean any mythological creature that is a composite of two or more animals. The earliest literary reference to a chimera is found in Homer's *Iliad*, where it is described as "a thing of immortal make, not human, lion fronted and snake behind, a goat in the middle, and snorting out the breath of the terrible flame of bright fire."[1]

In modern biology, a chimera is an organism composed of cells from different zygotes (fertilized eggs or early embryos) that have been fused. The resulting organism has a mixture of tissue with cells from at least two sources of nuclear DNA. Human chimeras can arise naturally during pregnancy. Two fertilized eggs in a woman's womb that spontaneously fuse during early gestation will grow to term as a chimera. Also, when two embryos are in the womb and cells from one embryo migrate to and are absorbed by the other, the result is a chimeric child. The chimeric child's organs have different sets of chromosomes: skin cells, kidney cells, and blood cells may not have identical nuclear DNA.

In maternity testing there are cases where a biological chimeric mother's DNA does not match that of her biological child when cells taken from certain parts of her body are compared to her child's cells. Forensic DNA testing can result in false negatives when comparing different cells of the same chimeric individual.

People go through life not knowing that they are chimeras. There are no dependable estimates of how many human chimeras are born each year. The rate of natural fraternal embryo fusion is not well documented.

Chimeras are distinguished from transgenic animals or hybrids, in which one or more foreign genes is added to an animal's single set of chromosomes. Chimeras have two sets of chromosomes in various tissues in their body.

Artificial chimeras can be created when scientists fuse two zygotes of the same species (intraspecies) or two zygotes from different species (interspecies). Some women with mitochondrial disease who wish to have a biological child have turned their eggs into mitochondrial chimeras. Their eggs with mutated mitochondrial cells are infused with healthy mitochondria from the cytoplasm of a donated egg, resulting in an egg with mitochondrial cells from two genomes, called heteroplasmy.

In 1953 the *British Medical Journal* reported a woman with blood containing two different blood types. A rare case of chimerism called tetragametic chimerism occurs through the fertilization of two eggs by two spermatozoa, followed by the fusion of the zygotes; the resulting child has intermingled cells from two distinct genomes.[2]

Chimeras can be artificially produced in research by transplanting embryonic cells from one organism onto the embryo of another, as in injecting mouse stem cells into mouse blastocysts. A chimera called a "geep" was produced in 1984 by combining embryos from a goat and a sheep. Nine years later, Chinese researchers fused human skin cells and dead rabbit eggs to create the first human chimeric embryo.[3] In England, after a debate in the House of Commons in May 2005, Parliament ruled that human-animal chimeras or hybrids created to harvest for stem cells must be destroyed within the first fourteen days.

There has been great interest among developmental biologists in transplanting human stem cells into the embryos of animals. According to a committee of the International Society of Stem Cell Research, "Scientists widely consider chimera studies to be indispensable for answering fundamental questions in stem cell and developmental biology. In stem cell research, human-to-animal chimera experiments typically involve the transfer of multipotent or pluripotent human stem cells into animals in embryonic, fetal, or postnatal stages of development to study stem cell behavior."[4] Human embryonic stem cells have been inserted into two-day-old chick embryos as a model system for studying the stem cells' in vivo development.[5] There have also been chimeras or hybrids constructed between humans and sheep, cows,[6] pigs, mice, and primates.

Alta Charo, who served on President Bill Clinton's bioethics committee, described the scientific and ethical value of human-animal chimeras. "If it were the case that you could take an enucleated cow egg and a human cell, and create an entity that functioned like a human embryo early on, so that you could get the stem cells, and then it decomposed later on, you would have evaded the problems of human embryo research, because you would not have destroyed a viable human embryo."[7]

One of the questions that bioethicists have grappled with is: At what point does the creation of animal-human chimeras and hybrids raise ethical questions?

■ ■ ■

Scene: Developmental stem cell biologist Hector Lamont has engaged in pathbreaking research on producing neural cells from stem cells and studying their function in the brains of animals. Sarah Tessman is a physical anthropologist and president of a nonprofit organization opposed to the creation of mammalian interspecies genetic hybrids with the acronym SHACH—Stop Human-Animal Chimeras and Hybrids. Dr. Franklin moderates a public symposium where the participants explore the ethics of creating human-animal chimeras and hybrids involving the use of stem cells, for research and therapeutic purposes.

FRANKLIN: I am Rebecca Franklin, your moderator tonight. I am a medical geneticist, stem cell scientist, and bioethicist. In the last couple of hundred years, humans have begun to play an important role in species survival, and more frequently in interspeciation—the term used to describe the process of recombining DNA to produce new species forms. With the discovery of recombinant DNA molecule technology, it is now possible to integrate genetic material from two species into the oocyte of a single organism.

One of our panelists is Professor Hector Lamont of Harvard University. Professor Lamont has created animal models with human DNA. Professor Sarah Tessman is adjunct professor of physical anthropology at the State University of New York and serves as president of an organization known as SHACH—Stop Human-Animal Chimeras and Hybrids.

Professor Lamont, can you give us some background on the history of transplanting human genes into nonhuman species?

LAMONT: Before the discovery of recombinant DNA, which gave us methods of cutting and splicing genes and transporting them across species, scientists used viruses as vectors for moving genes from one species to another. It was, by current standards, a very crude system. The revolution in molecular genetics has made gene transfer much more facile and precise.

FRANKLIN: What are some of the reasons for moving human genes into the genomes of other species?

LAMONT: There are two reasons, scientific and commercial. By moving human genes into bacteria, we can study the function of the genes in a simpler system, whose genetics is more fully understood. Similarly, by transporting human cells or genes into animals, we can study their function in a new environment and gain knowledge of the signals that cause the cells to survive and replicate and the genes to be expressed. Since we cannot do these kinds of experiments on humans, instead we do controlled experiments in animals with human genes.

FRANKLIN: What about the commercial applications?

LAMONT: Human genes are introduced into bacteria to produce a human protein. This can be done on a large scale. An example would be inserting the human insulin gene into bacteria and then producing large quantities of the protein for therapeutic purposes. Human genes are also inserted into the eggs of animals so that the adult animal will produce the protein, for example in its milk, which can then be purified and harvested. Companies have invested in transgenic goats with human tissue plasminogen activator (TPA), a human protein that dissolves blood clots. Transgenic cattle were created to produce milk containing particular human proteins that may help in the treatment of emphysema. There are many applications of transgenic animals that do not involve human genes. Drug companies have become interested in transgenic animals that have been made to contract a human disease. They can then test drugs on the animals before human trials.

FRANKLIN: Professor Tessman, how has the public responded to these historical examples of interspecies gene and cell transfer?

TESSMAN: The transfer of human genes into bacteria or viruses, other than for reasons of pathogenicity, has not been viewed as ethically problematic because these prokaryotic organisms are not sentient beings. But once research scientists or corporations integrate human genes or cells into animals, which possess neurological systems and brains, we must take account of producing hybrids or chimeric life forms that breach evolutionary species barriers. Because this is a slippery slope, our organization opposes all mixing of genes between humans and animals.

FRANKLIN: Professor Lamont, could you clarify the distinction between human-animal hybrids and human-animal chimeras?

LAMONT: In human-animal hybrids, human genes are introduced into the embryos of the animal. Every cell of the animal contains the added human genes. In a human-animal chimera, human cells are fused into an animal embryo. The animal will have a combination of two kinds of cells—animal and human cells in different tissues of its body. In other words, all cells in the hybrid are genetically identical, whereas the chimera consists of genetically distinct cell populations.

FRANKLIN: Dr. Tessman, what are some of your supporters' concerns?

TESSMAN: Biologists can quibble about the meaning of "species," a debate that has taken us from Aristotle to Darwin to Ernst Mayr. But there is no quibbling about who is human. The creation of novel beings that are part human and part nonhuman is sufficiently threatening to the social order that it should be prohibited.[8] This includes transplanting human brain cells into a mouse or primate or transferring human genes into a pig blastocyst in order to make a human kidney in a pig. We humans have no right to hybridize or chimerize ourselves with nonhuman animals, beyond our literary fantasies and mythical creatures. Canada has legislation that prohibits transferring human cells into nonhuman embryos as well as transferring nonhuman cells into human embryos.

FRANKLIN: Professor Lamont, Professor Tessman says that a blanket prohibition of transferring human genes into animals should be adopted. What is your response?

LAMONT: A blanket prohibition is too restrictive. We have used xenotransplants of pig valves for human heart patients, which by strict definition creates a chimera. But I do think there should be some moral boundaries. Also, a distinction should be made between research and commercial practices. Scientists have considered transplanting human brain stem cells into the brains of newborn mice.[9] They are trying to study how human neurons behave in an animal. Eventually this research could lead to stem cell cures for brain disorders like Alzheimer's disease. If there were a blanket ban on transplanting human cells and/or genes into animals, then these avenues of therapy would be lost.

FRANKLIN: Where do we find guidelines for setting the moral boundaries?

TESSMAN: Unlike in Canada, there is no U.S. law that prohibits or restricts animal-human chimeras. President Clinton was deeply troubled by the prospect of a human-cow embryo.[10] A Senate bill was introduced in 2005 titled the Human Chimera Prohibition Act, but it did not get passed into legislation.

LAMONT: There are nonmandatory guidelines issued by the National Academy of Sciences, which proposed embryonic stem cell research oversight committees at each institution to manage sensitive experiments. Their standards serve as a model for those adopted by the International Society for Stem Cell Research, which recommends that experiments in which human totipotent or pluripotent stem cells are implanted into animal embryos be monitored by an oversight mechanism. Also, the NIH does not allow funding for research in which human embryonic stem cells or human induced pluripotent stem cells are introduced into nonhuman primate blastocysts.

TESSMAN: The NAS indicated that scientists should be aware of how an animal's functioning would be affected when human genes were incorporated into the animal's germ line and whether some valued human characteristics might be exhibited, including physical appearance. They also noted that it is not currently possible to predict what properties will emerge when the chimeras or hybrids are created. This is why our organization believes we should adopt a precautionary approach by prohibiting human-to-animal genetic mixing. Moreover, these are only guidelines and have no force of regulation or law. They guarantee only that there will be a patchwork of different moral standards.

FRANKLIN: Professor Lamont, do you see the moral boundary for managing experiments involving human-animal chimeras?

LAMONT: There is a moral boundary. It is breached when the animal takes on appearances, behavioral traits, or cognitive properties of a human. If after neural cells are transplanted into a mouse brain, the mouse is still a mouse with a brain that contains some human cells, then we have not crossed a moral boundary. But if a mouse gains a level of cognition that transcends its species nature, we have crossed a line.

FRANKLIN: How would you know without doing the experiments and observing the mice?

LAMONT: You wouldn't. But certain outcomes are clearly near impossible. Even with extensive human cell transplants, you would not expect the mice to begin talking. When human cells, even embryonic-derived cells, have been transplanted into the mouse brain, their functions are dictated by the microenvironment in which they are introduced,[11] including the host cells or the extracellular matrix, rather than some internal drivers in the human embryonic stem cells.

FRANKLIN: Have there been attempts to use animals for harvesting human embryonic stem cells?

LAMONT: There have been proposals for transferring human stem cells into a nonhuman fetus and then harvesting the resulting tissue or organ to be transplanted to humans.[12] One group has patented a technique for generating embryos by fusing human nuclei with cow oocytes, and another group has fused human nuclei with rabbit oocytes to generate embryos as a source of stem cells. Some scientists have considered using pigs as donors because they have similar physiological properties as humans. Also, they are a nonendangered species, produce large litters, and do not have a preferred ethical status in society. But there are problems with xenotransplants in animals, including rejection and the risks of animal retroviruses infecting human cells. The dangers to patients from animal retroviruses could be great.

TESSMAN: The British have given conditional approval for human-animal constructs they call "cybrids." These are formed when the nucleus of a human somatic cell is implanted in an enucleated animal oocyte. The human nuclear DNA will be exposed to the animal's mitochondrial cells and cytoplasm. The resulting embryo contains the DNA of two species in separate regions of the cell.

FRANKLIN: What is the purpose of this biological construct?

TESSMAN: It is claimed that the creation of cybrids is a response to the scarcity of human eggs. The animal's oocyte minus its nuclear DNA allows the integrated human nuclear DNA to form into a blastocyst so that human pluripotent stem cells can be harvested. By strict definition, these constructs are not chimeras, because if they were brought to term the life form would not have two kinds of cells. However, its one type of cell would have animal mitochondrial DNA and human nuclear DNA.

FRANKLIN: I get it. From the words "cytoplasm" and "hybrid" they derive the neologism "cybrid." Are there limits to these cybrids?

TESSMAN: The Brits do not allow the cybrid embryo to develop beyond fourteen days. But other countries prohibit them as they do traditional human-animal chimeras.

FRANKLIN: What about experiments in which human cells are implanted into animal embryos? Wasn't there a group that planned to populate the entire mouse brain with human neural cells?

LAMONT: Yes, that was planned, but after the media and some bioethicists raised concerns, the experiment was not undertaken, or at least postponed. There was no unanimity among scientists about how to proceed. Some thought the experiment should go incrementally, taking measured steps while observing whether the human embryonic stem cells produced any human brain structure or human behavior. If such cases were found, the experiment would be aborted and the animals sacrificed.[13]

FRANKLIN: As I understand it, that was not the conclusion of the NAS. Professor Lamont, you reviewed and commented on early drafts of the NAS report. How would you characterize its conclusion?

LAMONT: In assessing the risks of transplanting human genes into an animal's gametes or embryo, the NAS minimized the probability that human embryonic stem cells or their derivatives would create an animal with distinctly human capacities or physical characteristics. It is reminiscent of the precautions taken by the National Institutes of Health during the recombinant DNA controversy of the 1970s.

FRANKLIN: At that time scientists organized an international meeting at Asilomar, California, to assess and manage the risks. Was anything like that done for chimeras?

LAMONT: No, but the NAS working group incorporated some of the principles formulated at Asilomar.

FRANKLIN: Wasn't the Asilomar conference designed to prevent the inadvertent creation and release of new pathogens from recombining genetic material in different species?

LAMONT: That's true, and even though the subject matter is different, the principles behind the risk analysis at Asilomar can be applied to the ethical risks of creating animals with human properties. For example, the closer the species is to humans, like primates, the greater the risk that the human-animal chimera will breach an ethical threshold. The NIH working group proscribed research in which human embryonic stem cells are introduced into nonhuman primate blastocysts.

FRANKLIN: What about the amount of hES cells introduced into an animal? One of the proposals called for replacing all the mouse's neural cells with human neural cells.

LAMONT: The NAS proposed going slowly by starting with small numbers of hES cells and determining whether they changed the structure or behavior of the mice. When no evidence of such effects appeared, investigators could increase the numbers. Before inserting hES cells in the blastocyst, they would start at a higher developmental stage, the embryo, and observe whether there were any untoward effects.

FRANKLIN: What would be the highest risk—the worst-case scenario?

LAMONT: The greatest risk is replacing the entire inner cell mass of a primate blastocyst with human embryonic stem cells.

FRANKLIN: Why is this such a high risk?

LAMONT: In this case you would have an inner cell mass composed entirely of human embryonic stem cell derivatives—pluripotent cells, surrounded by a nonhuman trophectoderm. There is a worst case chance that this chimera could be gestated to produce an organism with distinctive human traits and capacities.

TESSMAN: The approach taken by the majority of the scientific community is focused on outcome. They fail to see the repugnance of mixing the DNA at the outset. It is what Leon Kass referred to as "wisdom of repugnance" and what Arthur Caplan later abbreviated as Kass's "yuk" factor in human-animal interspeciation. The ethics should be rooted in the mixing itself, not in the consequences per se. Sheer repugnance can lead inexorably to the intuition that certain practices are ethically abhorrent.[14]

If you deal only with consequences, some judge or jury will have to determine whether the distinctive emergent properties in the chimera breach an ethical threshold. They would have to decide which features are associated with human dignity and which appearances are too close to human features.

FRANKLIN: Professor Tessman, don't you think it is possible to set up early warning signals to stop researchers from entering dangerous ethical ground? In that way society could benefit from a lot of valuable science that would stay within a safe ethical range.

TESSMAN: The NAS recommendations are just that—recommendations. They do not embody the force of law. Even if the federal government issued guidelines or regulations like the NIH guidelines for recombinant DNA research, they provide no oversight over the private sector or research done outside of U.S. borders. Currently, each investigative team and animal ethics committee determines where the ethical boundary resides.

FRANKLIN: Wouldn't it be sufficient to adopt a universal prohibition against breeding chimeras or hybrids?

LAMONT: That would be too restrictive, unless it was clearly an animal with human qualities. And if such a chimeric individual were produced, its life would be terminated. Of course, most countries that permit human-animal chimeras or hybrids do not allow the embryo to develop beyond fourteen days.

TESSMAN: In the absence of any early termination policy, once the human-animal chimera or hybrid is gestated—we are all aware of how many mice escape from laboratories. Once that happens, human control of the breeding is lost.

FRANKLIN: The current situation in the United States is somewhat paradoxical. The consensus is that scientists avoid creating human-animal chimeras or hybrids with human cognitive, behavioral, or phenotypic properties. If such a humanlike animal is created, on one hand, many believe it should not be allowed to develop; on the other, its human qualities suggest that the research subject should be treated with greater moral status, including perhaps protection, than the animal from which it was derived.[15]

DIALOGUE 19

STEM CELL TOURISM

The term "medical tourism" refers to patient travel from industrialized nations to foreign clinics for medical treatment. Patients seek foreign treatments for a number of reasons that include lower costs;[1] access to procedures, medical devices, or drugs that are unavailable in their home country; shorter waiting lists for treatments such as surgery; luxurious and private accommodations in foreign treatment centers; and assured privacy and confidentiality.[2] Under the U.S. regulatory system a drug must be proven safe and effective before it can be licensed for consumer use. The threshold for approval differs across national boundaries. Some nations allow physicians more discretion than does the United States in using experimental drugs. These physicians use their own criteria to evaluate the safety or effectiveness of a treatment.

Historically, Americans afflicted with cancer have sought treatments abroad when U.S. doctors have nothing left to offer them. Among the most publicized cases of cancer tourism in the United States took place in the 1970s, when patients traveled to Mexico for treatments with Laetrile, a substance found in apricot pits long touted as an antitumorigenic agent, but proven ineffective.

China attracted foreign tourists for medical treatments after its State Food and Drug Administration promoted Gendicine, the first approved drug for clinical use in gene therapy, in January 2004 to treat head and neck squamous cell carcinoma. Gendicine is a recombinant adenovirus with some genetic pieces removed and others added. Adenoviruses are a class of viruses that elicit respiratory, intestinal, and eye infections in humans. One of the main added components of Gendicine is the p53 gene—a so-called "tumor suppressor" gene, or what others have termed the "guardian of the genome" because it is known to express an oncoprotein that kills tumor cells. For years, p53 and its mutations have been studied for their relevance to the cause and prevention of cancer.[3] There is no evidence in the published medical literature that Gendicine works.

While we use the term "medical tourism" to describe the foreign travel of individuals seeking low-cost and effective treatments, it also applies to poor people traveling to hospitals in another, more medically advanced developing world country. "Narayna Hrudayalaya, a heart hospital in India, treats indigent people from neighboring countries—Pakistan, Bangladesh, Burma—who suffer from heart disease and cannot afford surgery. The treatment for them is free."[4]

When stem cells were widely touted as offering great promise for curing degenerative diseases, clinics applying unproven therapies opened up around the world in both industrialized and developing countries. Many cases of fraudulent stem cell treatments have been reported. A British medical doctor lost his license in 2010 for injecting patients with stem cells taken from cows. Other patients developed tumors from injections of stem cells. The *Harvard Gazette* reported, "There are clinics all around the world—but especially in China, India, the Caribbean, Latin America, and nations of the former Soviet Union—that will provide stem cell treatments for those long-intractable conditions. Never mind that cancer is the only disease category on that list for which there is published, scientifically valid evidence showing that stem cell therapy might help. Thousands, if not tens of thousands, of desperate people are flocking to clinics that charge tens of thousands of dollars for every unproven treatment."[5]

The next dialogue addresses the issue of the role of the state in protecting its citizens by prohibiting or warning against traveling abroad to receive uncertified experimental treatments with stem cells. Can and should states play the role of "medical nanny" for people intending to travel abroad, as they do in this country? Dr. Franklin moderates a heated exchange between two individuals who hold polarized views on stem cell tourism.

■ ■ ■

Scene: An open exchange of two contrasting views at the meeting of the International Stem Cell Organization. Dr. Robert Flossel is a stem cell scientist who works at a private nonprofit company and a self-avowed libertarian who advocates getting government off the backs of scientists and out of the business of regulating clinical trials. Dr. Barbara Grant has written a book arguing that there are insufficient ethical standards for clinical trials in Third World nations and wants Western nations to refuse data from trials that do not meet minimum standards.

FRANKLIN: Our exchange today is about the question of whether the activity known as "stem cell tourism" should be supported as the quintessential free market of frontier medical therapies or should be abolished as a cruel commercial abuse of medicine that gives patients false hope. For the purpose of this exchange of ideas, we define "stem cell tourism" as the activity of patients traveling outside of their own countries in search of unproven stem cell-based interventions to help improve or cure their illnesses. It is generally understood that at times, people travel outside their own country for medical treatments. Frequently, people who have resources and live in a place with underdeveloped medical care or technology travel to a country where they can receive the most advanced care. Recently, "health care tourism" describes Americans seeking less expensive surgery in advanced medical centers in Europe. In the case of stem cell tourism, people seek treatments that their own country will not provide because they have not been proven to the satisfaction of the scientific and medical communities.

This will not be a traditional debate. Instead, the two panelists will be allowed to converse with each other ad seriatim. Let's begin with Barbara Grant. Dr. Grant holds a Ph.D. in genetics and a master's degree in international relations. She recently published a book titled *Clinical Trials in the Third World: Discounted Patients.*

Dr. Grant, do you believe that stem cell tourism should be regulated?

GRANT: Desperate people with illnesses and nowhere to turn are preyed upon by clinics and physicians who offer them false hope. These individuals are being exploited as human subjects without any evidence that the treatments have any benefits. They enter treatments that, strictly speaking, are not clinical trials. The medical personnel are not accountable to institutional review boards or ethical standards that would protect patients from being made worse off or from the psychological trauma of having their hope raised, only to face the dire consequences of not being helped after paying considerable amounts for the treatment.

The only treatments the FDA recognizes using adult stem cells are for some blood disorders, skin regeneration,[6] corneal resurfacing,[7] and rare immune deficiencies.[8] FDA warns consumers that except for cord blood for certain specified indications, there are no approved stem cell products.[9]

FRANKLIN: Dr. Flossel, it has been reported that thousands of people are putting their health and their savings at risk to travel to private clinics around the world for unproven and potentially dangerous stem cell treatments. Doesn't something have to be done to protect these people from stem cell quackery?[10]

FLOSSEL: You are presuming that adults cannot decide what therapies they would like to try to improve their lives. So long as the medical treatments are legally permitted in the nation in which they are offered, the autonomy of the patient must prevail. Perhaps one out of ten patients will be helped. That is not a statistic that would get the therapy approved in the United States. But a person who has a life-threatening disease with low life expectancy may find it rational to take those odds. Why should a government agency prevent them from engaging with a clinic in another country that is working at medicine's frontier? The United States, after all, has restricted work on stem cells for political reasons.

GRANT: We are not talking about desperate people attending respected research hospitals or highly ranked clinics for experimental treatments. Most of these places are fly-by-night operations. There may be a physician involved. But in some of these countries, physicians can operate without a licensing framework. They work with nonphysicians such as biomedical engineers. The primary function of these operations is to instill false hope in a patient in exchange for his or her life savings.

FLOSSEL: I agree that there are some scam operations and that they should be closed down. We see this throughout medicine. But it is the responsibility of the country in which the "so-called" stem cell clinics are operating to shut them down or properly license them.

But what about the clinics that offer stem cell treatments that are known by medical authorities and have their imprimatur? They may be using techniques that are not approved in the United States or that are recognized as so experimental that they're not ready for clinical trial, but are not illegal in those countries. There may even be American doctors who are affiliated with the foreign clinics because they feel the U.S. policies are too restrictive.

I cannot see any reason to restrict a citizen of the United States from signing on to an unproven treatment in another country. After all, many people in the United States enroll themselves in *regulated* trials when the treatments are unproven.

GRANT: You're correct about our clinical trials. People are exposed to unproven therapies. Nevertheless, it is through the clinical trial process that evidence is produced, which eventually justifies the adoption of a therapy as a "standard of practice."

FRANKLIN: But clinical trials in the United States are carefully managed and monitored—certainly in comparison to how the trials are run in developing countries. There is an informed consent process, which requires complete disclosure of the risks and potential benefits of the trial intervention. Also, there is an independent institutional review board (IRB), which must approve the experimental treatment protocols and the informed consent process. If a trial is producing more harm than benefit, it is stopped before completion. If the drug benefits are unambiguous before the trial is completed, the placebo subjects are put on the medicine and the trial is terminated.

When a person attends a foreign clinic for stem cell treatment, he or she is not treated within a monitored and managed ethics review system, where transparency is mandated and an independent review of the risks is made. These patients are essentially offering themselves as human subjects for unregulated, perhaps unjustified, and potentially dangerous research for which they are paying. They are, in essence, taking part in a vast human experiment.[11]

FLOSSEL: Let's try to be concrete. Where would you begin to set restrictions? The XCell Center is a private clinic group and institute for regenerative medicine located in Düsseldorf and Cologne, Germany. They provide therapeutic use of autologous adult stem cells and perform medical stem cell research. They claim to offer patients with degenerative diseases the opportunity to undergo an innovative and promising stem cell treatment. And since their start in January 2007, they claim that more than 4,000 patients have safely undergone their various stem cell treatments.

They announce on their website that they are the first private clinic worldwide to hold an official license for the extraction and approval of stem cell material for autologous treatment.

FRANKLIN: They also have a clear disclaimer stating that "while stem cell therapy can help alleviate symptoms in many patients and slow or even reverse degenerative processes, at times it does not work for all patients. Consequently, the XCell Center can neither predict nor guarantee success for individual patients who undergo stem cell treatment."[12]

Is there anything wrong with patients from the United States going to Dusseldorf for this treatment?[13]

GRANT: The rules are different between the United States and Germany. It is unheard of in this country that a private company with licensed physicians would be providing stem cell therapies that have not been approved by the FDA, whether or not the therapies were part of a clinical trial. We cannot stop people from going overseas to obtain medical treatments that are not approved here. Cancer patients once streamed to Mexico to obtain Laetrile (Laetrile is a substance found naturally in the pits of apricots and various other fruits) to treat their cancers. Laetrile was never approved as a cancer drug in the United States and eventually was shown to be ineffective.

FRANKLIN: We can notify physicians to warn patients that stem cell tourism promises treatments that have not been approved in this country. Patients need to be informed that there is no peer-reviewed literature that supports the efficacy of many stem cell treatments, warned about the risks, and informed to ask many questions, such as: Is the clinic licensed by the government? Is the clinic insured? Have the physicians published any papers? Is the clinic monitored by a government agency? Are the physicians associated with a research university? Can prospective patients communicate directly with other patients who have received the treatment? Is there an independent ethics board to whom the clinic is accountable? These are the types of protections supported by the International Society for Stem Cell Research.[14]

FLOSSEL: People who are desperate cannot wait for clinical trials and do not want to be randomly chosen and placed into the placebo group. There is a clinic in San Jose, Costa Rica, that has treated as many as seventy multiple sclerosis patients with stem cells taken from fat tissue, with some remarkable success.

GRANT: The doctors administering the stem cell treatments charge as much as $70,000 per patient, and there has never been a clinical trial for this MS therapy. We cannot accept a patient's or a doctor's testimony about the success of the treatment. Many people claimed Laetrile was effective until a randomized double-blind clinical trial showed that it was not. We do not want stem cell therapy to follow the path of human gene therapy—test on humans before animals.

A majority of Internet sites that market directly to consumers falsely claim that stem cell therapy is ready for public consumption and is not experimental. Stem cell clinics overpromise the benefits of their treatments and grossly downplay or ignore their risks.[15]

FLOSSEL: There is a vast divide between clinical trials and preclinical trial evidence based on individual cases. Following your logic, governments would discourage innovation and personalized treatments. Remember what happened during the early AIDs epidemic. AIDs victims were purchasing drugs from foreign countries as a last resort. That helped some people have hope and live longer with the disease. That was well documented in the film *Dallas Buyers Club*.

GRANT: I am not for prohibiting responsible innovation, whether inside or outside the United States. But I would like to prohibit the quacks and the unethical stem cell treatments.

Stem cell treatments must have a rationale in the peer-reviewed literature, preclinical evidence of efficacy and safety, and animal data. Experiments without these safeguards should be banned by responsible medical authorities and international medical associations. There is still plenty of room for medical innovation outside of the random, double-blinded placebo clinical trial that affords desperate people a glimmer of hope and contributes to generalizable knowledge of medicine.[16] But we must take responsibility under the Hippocratic Oath to "do no harm." How can we do nothing when we hear about the Israeli boy who received stem cell treatments in Russia for spinal cord injury and developed multiple tumors or the woman treated in Thailand for the autoimmune disease lupus, who, after stem cell treatment, developed kidney failure and died from sepsis?[17]

FLOSSEL: The difference between us, Ms. Grant, is that I take a libertarian stance on what people choose to do about their health care, while you believe that institutions, national or international, should set conditions that limit our choices.

GRANT: That's precisely what we do when we limit the right to dispense medical treatment to those with a medical license and restrict the use of drugs to those approved by the FDA. We need the same protections for stem cells.

FRANKLIN: Liberty is a fine concept until individual decisions affect the community. Does the state or international community have an obligation to protect people from their autonomous but short-sighted decisions when they fall prey to the purveyors of false hope who profit from voodoo medicine? Government's primary function is to protect the health and well-being of the citizenry from unscrupulous deception. Stem cell tourism is challenging this premise.

DIALOGUE 20

SOCIAL MEDIA MEET SCIENCE HYPE

Science and the media have become irrevocably intertwined. The mass media report scientific results in leading journals in a form accessible to the popular reader. Increasingly, this summary reporting exaggerates the significance of the results. For example, imagine a study reporting that mice, after being administered a drug, learn to run a maze more rapidly than controls. Once the scientific publication passes through the media food chain, it will most assuredly be announced by a headline, "Drug Enhances Intelligence." All the caveats and limitations scientists offered in the journal article will be ignored in the popular version.

The term "hype" has a pejorative connotation. It has been defined as an action, written or verbal, to promote or publicize a product or idea, often exaggerating its importance or benefit. Specifically, hyping science is associated with "overemphasizing benefits, underplaying risks, and making grandiose claims and promises [raising] the hopes and expectations of individuals."[1] Hype is about enhancing the importance of a study beyond what the data reveal, calling attention to possible future pathbreaking applications in medicine.

However perceived, hype has a distinct function in science. First, it alerts the public that something of note has taken place in the laboratory. The hyped story prepares consumers for the next generation of medical treatments. If the headline is about a new drug trial, consumers might ask their health provider about it. Even if the drug has not yet been approved by the FDA, the hyped story may encourage some patients to inquire whether they can receive an off-label prescription or enter an experimental trial.

Second, the exaggerated claims of a study can position scientists in a competitive field to get more funding for their work. The media attention may be viewed by those granting the funding as validation that the work accomplished by the scientific team is important and especially noteworthy, and seen that way by the journal and the general public. "Because all researchers are competing for a slice of the same funding pie, a given researcher or institution has little to gain by mediating the hyperbole."[2]

Also feeding the trend toward science hype are the journals themselves. The top journals—those with the highest citation rankings (impact factor) and the highest rejection rates—cater to the media. They offer early release of articles to science writers with an embargo attached, which builds suspense. Journalists tend to believe that the articles published in the high-profile journals like the *New England Journal of Medicine* (NEJM), the *Journal of the American Medical Association* (JAMA), *Nature*, *Science*, *The Lancet*, and the *British Medical Journal* are worthy of the most publicity because they were selected for publication according to their impact on the field.

Public relations officers at universities are mandated to give high visibility and publicity to scientific breakthroughs by their researchers, which they highlight on their websites and in alumni magazines to foster pride and stimulate donors. "Media coverage brings attention to their research, helps attract funding and raises the profile of the institution."[3] An editor in *Nature Genetics* wrote that the "copious press releases emitted by universities, research institutes, granting agencies and journals" are one potential source of science hype.[4]

Third, science hype helps scientists partner with venture capital investors who are seeking early investment blockbusters. Caulfield notes that "the hyping of research results might be part of a more systemic problem associated with the increasing commercial nature of the research environment."[5]

The entire system—university, researcher, journal, government or private funder, the media, and the investor—is self-reinforcing for producing hyperbole about scientific results. "The job of the PR people is to get press coverage of science, but they experience the same pressure that scientists do when writing their grants—to link their work as directly as possible to human disease or to hype the work beyond its real scientific meaning."[6]

The negative side of hype cannot be ignored. Exaggerating the importance of scientific breakthroughs can distort research budgets, lower the public's confidence in science, create simplistic and overly optimistic expectations about what it takes for scientific results to produce concrete benefits, and undermine the public's understanding of how science actually operates—with small baby steps that meet impasse after impasse before a useful result emerges. "Media stories that emphasize unrealistic, near future benefits will inevitably result in unmet expectations."[7] Perhaps one of the worst consequences of hype is the premature translation of research into clinical trials with a deadly outcome. Hype can sometimes cause regulators or clinicians to skip steps in transferring the science to medical application.[8]

In the next dialogue, a medical anthropologist who understands the culture of science has a frank discussion with stem cell scientists about whether stem cell discoveries have fallen prey to the "science hypesters." The overreach of basic science into premature claims of medical cures is a manifestation of the tensions among human aspiration, wish fulfillment, and fallibility. The real scientists discussed in this dialogue who devoted their research lives to the pursuit of medical cures, regardless of the results, should be viewed as people of remarkable tenacity and creativity following their inner voice on a path toward the improvement of the human condition.

■ ■ ■

Scene: Medical anthropologist Ann Cummings of the University of New South Wales discusses the role of hype in science with stem cell biologist Bretton Salisbury of Duke University, who does research on and has a small startup company that will be developing therapeutic uses of stem cells. Dr. Franklin has consulted with the company on their project to develop stem cells for spinal cord injuries.

CUMMINGS: Before I begin asking you some questions about your stem cell institute at Duke, let me tell you something about my background. As a medical anthropologist, I am interested in how a field of medical research evolves from a particular culture and how the practice of medicine reflects cultural norms. With respect to your work on stem cells, I am interested in understanding social responses, cross-culturally, to discoveries and claims about stem cells. My first question to you both is: How did you become involved in stem cell biology?

SALISBURY: I had the privilege of being a postdoc in the laboratory of James Thomson at the University of Wisconsin. In November 1998 our group derived embryonic stem cells from human blastocysts. For the first time I experienced the heightened exhilaration that comes from scientific discovery. Not only had the discovery been a grand slam in the scientific community, but also the news went viral in the media.

FRANKLIN: I was a medical geneticist when my father became a quadriplegic from a fall. I decided to devote my life to finding a cure for his paralysis through stem cells.

CUMMINGS: I followed the news coverage after the Thomson team's discovery was published in the journal *Science*. The *Times of London* ran the headline, "Transplant Teams Grow Cells from Embryos." According to these stories, "stem cells" meant organ transplants. The *Times of London* wrote, "Scientists have taken an important step towards growing organs in a laboratory for use in transplantation." How did you feel about the media's enormous leap from stem cells to organ transplants?[9]

SALISBURY: You are right about the media. They took a single important discovery and fast forwarded it over a thousand steps, any one of which, if unrealized, could derail their prediction. And yet, while the media take leaps of imagination, so do scientists when we project a basic discovery to a medical miracle. The thought of what the discovery could potentially do elevates its importance to the entire community.

CUMMINGS: Distorted expectations of science could backfire. It wasn't just the London papers that exaggerated claims about stem cells. *The New York Times* is not known to rush to judgment about medical discoveries. Yet, a *Times* science writer wrote, "scientists for the first time picked out and cultivated the primordial human cells from which an entire individual is created."[10] Did you think that the cells you had isolated were capable of making an entire individual?

SALISBURY: We felt we had discovered an important piece in the puzzle of development—how a single cell is the precursor for over two hundred cells in the human body. The cell you start with, the zygote, undergoes the first division into a two-celled embryo, and after about five days the blastocyst is formed. We needed to extract and culture the inner cell mass of the blastocyst as the source of embryonic stem cells before they differentiated further and disappeared.

CUMMINGS: Discovering a key piece of the scientific puzzle of development is a far cry from promising, even if only by implication, cures for major diseases. Do you feel that scientists are enablers in the process of gross extrapolation from discovery to cures?

SALISBURY: Now you are entering a gray zone of scientific ethics. Those of us at the frontiers of medical discovery can envision a future that few others can understand. But a single medical discovery cannot make it happen. We have to create a buzz of expectation.

CUMMINGS: As a medical anthropologist, I would use the term "hype." Are you saying that hyping science has a special function in your work?

FRANKLIN: I can answer that with brash honesty: Absolutely! No field will prosper without some level of hype. Most of the predictions will never see the light of day. But breaking through new fields of medical research requires resources, to which hype contributes, and the exaggeration is worth it if it leads to even one major medical therapy. Would we have mobilized public and private resources for sequencing the human genome without exaggerated claims about personalized medicine? It certainly would have taken years or decades longer. The hype of science mobilizes a new generation of students to enter the field. It also provides the incentives for postdocs to spend sixteen hours a day in the lab in the hope of making the next big discovery for a cure. In my case, I needed to convince myself that this research was going to save my father from living the way he currently does.

CUMMINGS: What about false hopes of people who believe that medicine is truly on the cusp of curing the disease of a loved one? Is it fair to build up and fail to meet these hopes? It's likely that many of the current sufferers from conditions for which stem cell therapies have been promised will be long dead before the therapies actually arrive, if indeed they do.[11]

SALISBURY: The problem is that you don't know it's false until you put all the resources and the knowledge into carrying forward the program. Had we worried about false hopes, we never would have had a kidney transplant, a procedure that was once highly problematic and is now routine.

CUMMINGS: I do see your point about taking risks in science and medicine. But there is something quite disingenuous about exploiting the fears and expectations of people who have little hope in order to build resources for a new area of study. Consider what happened with human gene therapy. When it was initially promoted and the public and political resistance overcome, gene therapy was touted as a treatment for the unfortunate people suffering from a genetic disorder. It was put forward as the only hope of a cure or disease mitigation. Today, however, we know that most gene therapy projects are not directed toward genetic diseases but toward the treatment of common diseases. The patient groups that were exploited to gain political and public acceptance of gene therapy have not benefited significantly from it, and many of the people afflicted with rarer forms of genetic disorders are unlikely ever to benefit.[12]

SALISBURY: I thought briefly about working in human gene therapy, but I changed course after I realized it wasn't heading where I hoped it would.

CUMMINGS: What do you mean?

SALISBURY: I thought it was promising for treating Mendelian (single gene) disorders. But then I was shaken by the death of young Jesse Gelsinger in 1999 at the University of Pennsylvania. He was one of the earliest human subjects for gene therapy trials,[13] and he died a few days after the initiation of his experimental trial, probably from his body's reaction to the virus that was used to deliver the corrective gene into his cells. It seemed too risky to me.[14]

CUMMINGS: Undoubtedly you recall the excitement and the hype around human gene therapy. Some disease communities became quite excited about the promise and lobbied for greater resources. By the new millennium there were 516 big stories in the media and nearly 2,000 federal grants; and by 2003 there were nearly 1,600 citations in the scientific publication database Medline and nearly 600 gene therapy clinical trials. The lure of profits from human gene therapy had driven the formation of more than 160 companies. The hype did exactly what you claimed it does, namely, activate and consolidate resources into the field. But it also raised false hopes. One of the negative outcomes is that it caused patients to neglect other potentially effective treatment options.[15]

SALISBURY: I trust your data. If science has overstated its case, it will correct itself. We will see resources redirected to other, more promising areas. Industry will withdraw funding if there is no return or little hope for return on the investment in five years. Scientists will retrain in other subfields, like stem cells, as I did.

FRANKLIN: I was perfectly happy in medical genetics before I entered stem cell science. That field had no fewer hypesters than my current field. They spoke about the personalized genome transforming medicine. Under this scenario, each of us would carry around our sequenced genome that would be our "personal Holy Grail" for medical treatment.

SALISBURY: This may seem like fantasy today, but in twenty years it will be part of our medical records. There are already important parts of our genome that tell us about our tolerance for certain drugs.

CUMMINGS: But there is an industrial side to all of this. When you're dealing with venture capitalists and corporate boards, they don't turn on a dime and forget their investment. They want payback. This means pushing the envelope and finding applications for human gene therapy that are questionable. Some people feel that Jesse Gelsinger should not have been a human subject and perhaps he was enrolled prematurely, before there were good animal experiments, because there was a commercial interest by the principal investigator and the University of Pennsylvania.

SALISBURY: Gene therapy has had a run for three decades. We have to give the same opportunity to stem cells. The results are beginning to appear. Scientists have begun to use embryonic stem cells to make human skin cells, bone cells, liver cells, muscle cells, tendon cells, and neurons (dopaminergic and motor). We are seeing breakthroughs in cell and tissue engineering that have surpassed the potential of gene therapy.

FRANKLIN: Science cannot avoid entering public relations, because the individual fields compete for social resources. Stem cell funding is in competition for both public and private investments with other important areas of new research—neurobiology, nanotechnology, synthetic biology, and the personalized genome, to name a few. Public relations will always be part of the competition among the medical disciplines for public resources and private investments. Success, however, will not be determined by the hype, but by the science.

CUMMINGS: There is a side to hype you are not understating. Because of the exaggerated and premature claims, stem cell hucksters are capitalizing on people's false hopes. Stem cell tourism has become a profitable scam. And it's fed by all the stem cell hyperbole that cures are around the corner.[16]

SALISBURY: To be honest, we have had decades and decades of hype in cancer research. Every time a new antitumor chemical is developed, there is a tsunami of excitement. This pumps investment and drives research to the next level. Venture capitalists are very mercurial. They have a three-to-five-year attention span. In terms of the arc of scientific research, we consider that ADHD-type behavior. When they read that a new molecule with a botulinum toxin targets tumor cells, they will make a mad dash for investment. Invariably, the magic bullet expectation for curing cancer rises and falls. But even without the magic cure, there are many cancer drugs on the market that have prolonged lives.

CUMMINGS: We shouldn't forget the excitement over angiogenesis. It seemed like an amazing breakthrough that had such an appealing narrative. Tumors need a blood supply to grow. Angiostatins are proteins that block the blood to tumors. Ergo, cancer is cured.

FRANKLIN: Granted, the social media saw enormous potential in angiostatins—at least initially—reflecting the view of a segment of the scientific community. One cannot deny that the high expectations for this class of drugs were never realized, but as so often happens, the creative spirit of science finds other uses. Some angiostatins have been approved for use in wet macular degeneration, where capillaries leak into and destroy the macula in the eye's retina and cause a loss of vision.

CUMMINGS: This is all fine, but where does it leave the unfulfilled hopes of the cancer patients betting on a cure? Remember the words of Judah Folkman, who spent his entire career developing the theory of tumor angiogenesis: "If you have cancer and you are a mouse, we can take good care of you."[17]

DIALOGUE 21

FEMINISM AND THE COMMERCIALIZATION OF HUMAN EGGS/EMBRYOS

The next dialogue, which takes place between two feminists separated in age by a generation, explores how women relate to the moral status of embryos. Feminism can be described as a series of social and political movements committed to furthering the rights, opportunities, self-identity, and power relations of women. The feminist movements are viewed by historians and gender studies scholars as having evolved through three stages, sometimes referred to as waves. The first wave of feminism developed in the United Kingdom and the United States during the nineteenth and early twentieth centuries. Women organized around property rights, voting rights, and the right to engage in legal contracts. First-wave feminism was a human rights movement. Women worked, raised families, and paid taxes, yet they were restricted from owning property or voting. First-wave feminists identified with and supported the antislavery movement, seeing that both slaves and women were treated not as citizens but as property. When the Nineteenth Amendment to the Constitution was passed in 1920, giving women full voting rights, the suffragette movement, which had organized around getting women the right to vote, dissolved, having fulfilled its goal. The first-wave feminists were largely middle-class, educated white women.

Some years later a new, more diversified feminist movement was born. The second wave of feminism began in the 1960s and focused on reproductive rights, self-help, self-determination, sexual power, consciousness raising, and economic opportunity for women—that is, breaking the glass ceiling of employment. Women in this movement helped pass the Equal Rights Amendment. The National Organization of Women was founded in 1966 and was devoted to women's equality. Second-wave feminists provided a large tent for radical as well as moderate feminist groups that brought public attention to sexism, the objectification of women, patriarchy, and the commodification of sexuality in the media.

The development of the birth control pill in conjunction with the Supreme Court decision in *Roe v. Wade* on abortion rights afforded women both technological and legal power to control their reproduction. With the right to have a legal abortion in the United States settled by the highest court, women finally gained control over their bodies, including a gestating embryo or fetus, at least in the first trimester of pregnancy. The personal lives of women were seen through a political lens that brought into focus unjust and deficient health care, economic disparities, and patriarchy, all resulting from structural inequalities in the distribution of wealth and power.

Second-wave feminists struggled to keep government away from their embryos. They rejected the idea that embryos possessed a value outside and independent of a woman's body. As noted by Deckha: "many feminists [argue] against the rise of visual technologies that encourage the public to view the fetus as a free-floating entity separated from a woman's body."[1] The subjugation of women was connected to broader critiques of capitalism, patriarchy, consumerism, and the despoliation of the environment. Feminism between 1960 and 1980 broadened to include working-class women, minorities, and Third World women. The ecofeminist branch of the second wave connected the abuse of nature with misogyny and argued that on biological grounds, women had a closer relationship with the natural world.

Third-wave feminism brought a departure from doctrinaire values and ideology. Women who identified with the third wave challenged essentialist definitions of femininity provided by their middle-class predecessors. They focused on defining feminism and opening it up to diverse perspectives that included beauty contest, high-heeled, and lipstick feminism. Women in the third wave emphasized empowerment over sisterhood. They transcended the stereotypes of the 1960s "women's libbers," redefining femininity and feminine beauty for themselves. Post-1990s feminists invested in an exploration of gender definitions and gender identity. They abandoned party line answers to gender and sexuality, even if doing so reinforced female sexual stereotypes.

How are the pro-choice values of the second-wave feminists reflected in the ideas of third-wave feminists who "do not deny the beingness of fetuses and embryos or rule out the application of 'respect' and 'dignity' concepts to them but yet still reach pro-choice conclusions"?[2] Can we place a moral value on the fetus that is being aborted? Will the "perceived concern about embryos" among the third wave "cede territory to anti-choice forces," turning back the clock after their forebears fought the battle over abortion rights?[3]

■ ■ ■

Scene: Dr. Franklin considers herself a second-wave feminist. Her grandmother, Sophie Franklin, was a suffragette in the early twentieth century and started a national women's organization. Dr. Franklin has always supported a woman's right to an abortion. Jenny Gunderson is a twenty-five-year-old activist with a women's health group that is devoted to protecting women from "eggsploitation," or the commercialization of their eggs, and its impact on women's health. They differ in their opinions about how feminists should relate to embryos.

FRANKLIN: Jenny, your organization has issued a very strong condemnation of embryonic stem cell research. Can you explain how you reached that decision?

GUNDERSON: Well, as you know, we have been active over the past decade in educating women about artificial ovulation and the growth of an industry that produces, sells, and distributes human oocytes. Since 1998, women's eggs have been intensely commercialized. The growth of research on embryonic stem cells (ESCs) has created a rapidly rising demand for embryos and human eggs. That demand is putting people at risk. Therefore, we support the continued passage of the Dickey-Wicker Amendment and its extension to private funding sources.

FRANKLIN: Your literature refers to a day-old embryo as "sacred life," "proto-human," and "ensouled plasm." Can you see that your organization is seeking to turn back fifty years of progress for women?

GUNDERSON: How so?

FRANKLIN: The Supreme Court ruling in *Roe v. Wade* in 1973 is premised on two ideas: first, women have autonomy over their bodies; second, the early embryo does not represent a sacred independent life—a person with rights that preempt the autonomy of the pregnant woman. By elevating the status of the early embryo in your literature to a proto-human, you are helping the right-to-life anti-abortion organizations. Your rhetoric provides grist for those antifeminist forces that want to take away a woman's right to an abortion.

GUNDERSON: The primary mission of our organization is to prevent commercial forces from controlling women's eggs. Poor women will sell their eggs in order to survive, regardless of how injurious it is for them to endure superovulation. When done for IVF, superovulation is a risk many women will accept in order to have a baby. When women do this for commercial purposes, the care they get is substandard and the number of ovulations they go through is much greater than for a woman seeking to become pregnant. The new demand for research oocytes and embryos from which to generate and extract embryonic stem cells has raised the stakes for women's health. We decided that the only way to eliminate this demand is to partner with those groups that find research with embryos in a dish unethical. What we say about research embryos is not relevant to embryos formed naturally in a woman's body.

FRANKLIN: There are many ways to advocate against research on embryos without treating early embryos as proto-persons, when all they are is a bundle of cells, with no consciousness and no feelings—in other words, no sentience. If you treat the early embryos as proto-persons or homunculi, you will turn back the clock to the time there were coat hanger abortions.

GUNDERSON: There are only a few viable strategies open to us for stopping the commercialization of eggs and embryos for research. We can argue that it should be banned because it will lead to the use of older and older embryos—the slippery slope. Or we could argue that it is unethical to use human life for advancing human life—even if the life is in its early development stage. We didn't think any of these arguments would be as effective as simply stating that early embryos represent sacred human life and should be protected against their use for research.

FRANKLIN: But by choosing that path, you are aligning yourselves with anti-abortion groups and contributing to their movement.

GUNDERSON: If women are allowed to sell their eggs for research, a market in eggs will emerge, valuing their reproductive tissue over their well-being. [4]We'll stop our campaign as soon as scientists stop extracting stem cells from embryos. But right now we are choosing the lesser of two evils.

FRANKLIN: In the United States, women are permitted to sell their eggs to some anonymous person who wants to have a child the egg donor will never know. They go through the same procedure as a woman would go through when she donates eggs for research. How can you conclude that providing eggs for reproduction is less exploitative or dangerous than providing them for research?

GUNDERSON: You are right. It is not less exploitative or dangerous. Our organization would like to stop all payments for eggs and embryos, period. Eggs and embryos should be treated like donated organs or tissue. A fertile woman who wishes to help an infertile woman give birth should be able to donate an egg, but without getting money for it. As soon as money is exchanged, the doors of exploitation are opened.

FRANKLIN: For some women, selling their eggs may be their only way of getting out of poverty or avoiding the burden of big college loans. Increasingly, feminist bioethicists, who advocate a woman's rights to choose, support women's right to be compensated for egg donation for reproduction or research.[5] If men can sell their sperm, why can't women sell their eggs?

GUNDERSON: When modern capitalism wants to set aside a public health or social justice framework that gets in the way of maximizing profit, they always appeal to libertarian principles. And you second wavers and some young feminists buy into it.

FRANKLIN: But isn't that what third-wave feminism is about? Let informed women make their own choices while understanding their rights and benefits. I do not agree with all decisions of third wavers, such as when they become complicit enablers of sexism. But I consider the goals of science to be part of an honorable enterprise. A woman who donates her eggs for research, paid or unpaid, is engaging in a noble act, not one that will diminish the value of women.

GUNDERSON: We believe that women must be saved from being enslaved to a commercial egg market. There is no autonomy in slavery. Just consider what the drugs for hyperstimulation of the ovaries for extracting eggs do to women: cause chest pain, nausea, depression, dimness of vision, loss of pituitary function, hypertension, asthma, edema, and abnormal liver function. It is no joke.[6]

FRANKLIN: By stopping the sale of eggs, you may be inadvertently contributing to a black market that brings egg prices to their highest level ever.

GUNDERSON: Our goal is to eliminate the research demand for eggs by making it illegal to use embryos in research. There is no black market when there is no demand. Scientists will, by and large, follow the law.

FRANKLIN: Threats to a woman's right to an abortion are the ultimate evil. Suppose our government made it illegal to sell eggs for research. Eggs would be treated like kidneys. You can donate a kidney, but you cannot sell it. The National Academy of Sciences proposed that women should not be compensated for donating eggs for research. Wouldn't that eliminate the problem your group is trying to resolve? And you could do it without elevating the moral status of or sanctifying the early embryo.

GUNDERSON: If egg donors were not compensated, then, I agree, the women we are trying to protect could not be exploited for their eggs. There would be no financial incentive for them to give up their eggs and undergo the treatments. But this would have to be accomplished internationally, because there could be a black market for eggs in Third World countries. I believe there is a greater chance of a worldwide ban on embryo research than a ban on selling eggs, since IVF eggs already are commercialized.

FRANKLIN: A Rand study estimated that there are over 400,000 embryos frozen. Many of these will be discarded and can be donated outside of a commercial market. Using these embryos for research does not exploit women since their eggs were obtained and fertilized for IVF purposes.[7]

GUNDERSON: Only a small percentage of the frozen embryos, about 2 percent, are designated to be discarded. Those are usually of lower quality. They will not meet the demand for embryonic stem cell lines, and thus there will still be a strong market for eggs and embryos.

FRANKLIN: The UK has given conditional approval for harvesting human embryonic stem cells by transferring human nuclear DNA into enucleated animal oocytes. Is that consistent with your advocacy?

GUNDERSON: If they can derive human embryonic stem cells without a human egg, they have our blessing.

FRANKLIN: Alternatively, Japanese scientists have created viable reproductive mouse eggs from induced pluripotent stem cells. If human embryos were made through this technique, would you have any opposition?

GUNDERSON: Human stem cells used for research that are derived from reprogrammed adult cells would not exploit women—as long as natural oocytes are not part of the process.[8]

FRANKLIN: These scientific advances may be our common ground.

DIALOGUE 22

WAS MY BIRTH EMBRYO ME?

How do we acquire a self-identity? When does it appear in children? And how do we connect our self-identity with the many states and stages of our physical and psychological development? This issue has occupied some of the greatest philosophical minds. A number of philosophers have linked self-identity to our brain or mental states. The distinguished British empiricist John Locke included a chapter on personal identity in his groundbreaking work *An Essay Concerning Human Understanding*. Locke, like many intellectuals of his time, believed in mind-body dualism—that two fundamentally distinct substances make up a person. Our identity, according to Locke, is linked to the human mind and the continuity of its memory states:

> For, since consciousness always accompanies thinking, and it is that which makes everyone to be what he calls self, and thereby distinguishes himself from all other thinking things, in this alone consists personal identity, i.e., the sameness of a rational being: and so far as this consciousness can be extended backwards to any past action or thought, so far reaches the identify of that person.[1]

The body changes significantly throughout one's lifetime. There is no continuity of the cells, since they die off and new ones get formed. If the same consciousness were moved from one body to another, personal identity would persist. "For, it being the same consciousness that makes a man be himself to himself, personal identity depends on that only, whether it be annexed solely to one individual substance, or it can be contained in a succession of several substances."[2]

Locke anticipated a thought experiment in which a person's brain is transplanted into another body. Most would concede that the self-identity of the individual follows the brain. The public identity, however, involving photographs and fingerprints, follows the body. The genetic identity of the individual most likely follows the body, where most of its DNA resides, with the exception of the brain cells. The genetic identity of an individual is determined from the unique DNA code of three billion base pairs. Only identical twins or clones could share the exact DNA code. But the self-identity of twins or clones would never be the same.

Our self-identity—who we believe we are—is the result of our memories of our life, our psychological continuity, how other people treat us or perceive us, and what we see when we look at ourselves in the mirror or in photographs. Our self-identity and our public identity can change dramatically throughout our lives, while our genetic identity remains relatively constant.

People may lose their self-identity when their brain has been traumatized. Those who have experienced complete memory loss must create a new self-identity with the help of those with whom they interact, unless their memories return. Amnesia victims retain their public and genetic identities.

If our self-identity is largely dependent on our memory states, stored in the brain's complex architecture, what happens if we can only remember so far back? If I cannot remember anything before age four, do I exclude any picture of me as a one-year-old as part of my self-identity—even if this picture was given to me by my mother, who told me stories about what I was like at that age?

Self-identity incorporates pictures and stories about ourselves that are not retrievable in our memory states from direct experiences. Our memories include the pictures and stories about us told by others, and they too contribute to self-identity. Some people's desire to complete their self-identity beyond their memory states takes them to a journey as far back as they can get: their ancestry. Adopted children seek their birth parents. The drive to complete the empty margins of their self-identity goes beyond their earliest memories to the genetics of their biological parents.

Memories, woven together in a disjointed tapestry, may be the most significant internal part of our self-identity. But memories can be false. For example, people have had false memories of being sexually abused. There is a psychological and legal literature that discusses false memory, the causes of which are complex. Sometimes false memories persist despite glaring inconsistencies and lack of validation. Once memories, whether true or false, become fixed in a person's psyche, they contribute to his or her self-identity.

The drive to know who we are and where we came from may motivate collecting photography, writings, or other objects of meaning. These collected artifacts reignite our punctuated memories that form the fabric of our self-identify. All of these external and internal factors form the collective subjectivity of who we are.

New technologies can push the limits of self-identity. Parents these days save pictures and videos of the pregnant mother's fetal sonograms. Someday these will be presented to the grown child. If a child was born by in vitro fertilization, then the picture of the in vitro fertilized egg prior to implantation is the source point of their being. The fact that people preserve these visual artifacts of their prenatal and early postnatal life reflects the value these objects have for self-identity, even if they reveal nothing about the phenotype or behavior of the individual, which changes so quickly and dramatically from the fetus and infancy to adulthood.

Can people feel some affinity and self-identification with their fetus when at most all they have is a sonogram image? In his essay, "Was I Ever a Fetus?" Eric Olson describes the "Standard View of personal identity and the fetus."

> The Standard View of personal identity says that someone who exists now can exist at another time only if there is continuity of her mental contents or capacities. But no person is psychologically continuous with a fetus, for a fetus, at least early in its career, has no mental features at all. So the Standard View entails that no person was ever a fetus—contrary to the popular assumption that an unthinking fetus is a potential person.[3]

Does a person's self-identity and connection with his or her fetus of origin follow a biological or a psychological paradigm? The answer will help to clarify the question for this next dialogue. An affirmative answer may explain why some individuals attribute moral status to a fertilized human oocyte.

■ ■ ■

Scene: Neuropsychologist Jacob Spencer discusses the concept of human identity with French philosopher-phenomenologist and metaphysician Antoinette Picard. The issue is when human identity emerges in development and whether human identity encompasses the one-celled embryo from which a person develops. Dr. Franklin spent a semester in Paris studying with Dr. Picard and was invited to the discussion in the office of Dr. Spencer at New York University.

SPENCER: Welcome, Antoinette and Rebecca. I understand that you spent some time together in Paris. When I was a young man I studied with François Jacob, and he inspired me to study the brain—the new frontier. Dr. Franklin, you organized this meeting to discuss the concept of identity in human development. Can you explain where your interest comes from?

FRANKLIN: I have been on a quest to understand the full scientific and ethical dimensions of stem cells, to which I am devoting my scientific career. The use of embryonic stem cells has raised questions about human identity. If we relate our identity to the earliest cells of our development—the zygote—then there is a natural transfer of moral status from who we are to who or what we came from. Therefore, I would like to ask you whether our identity can be traced to the primordial cells of our being.

SPENCER: Okay, now I get it. Antoinette, as a metaphysician, how do you approach this issue?

PICARD: I begin with several first principles on human self-identity. First, personal identity is determined by the continuous states of memory that we can recall to consciousness and the central ego that connects these states. Immanuel Kant had a fancy term for it: "the transcendental unity of apperception," or the organization of perception that occurs in consciousness. When we wake in the morning, we are aware that we are the same consciousness with the same memories who went to bed the evening before.

Second, personal identity is reinforced through family and community, without which our identity would be vacuously solipsistic. Finally, our identity is also shaped by our genome, our epigenome, and their environmental interactions.

SPENCER: Can a living organism possess memory states without having an ego or a self-identity? Certainly, machines can have memory states, which can be recalled without a self-identity connecting those states. But I defer to Antoinette about living organisms.

PICARD: I believe there is evidence that certain organisms possess primitive memory states without consciousness. They can learn from experience to avoid certain stimuli. But there is no evidence that they have consciousness or self-identity.

SPENCER: If the human organism develops from a fertilized egg to an infant, child, and then adult, identity cannot be a quality that exists at every stage of development. It must therefore arise at some point.

FRANKLIN: You are saying this because consciousness is a prerequisite for personal identity?

SPENCER: Precisely!

PICARD: I too do not believe there can be self-identity without consciousness and self-consciousness. The organism must be aware of a unity of self.

SPENCER: So when does identity appear?

PICARD: One-celled organisms do not have the biological capacity to store memories. There is not sufficient complexity—no neurons, no brain. The one-celled egg develops into an embryo and then a fetus and will eventually exhibit consciousness, and then postnatally, identity. There is no precise point in the development of the human organism.

SPENCER: The question before us is whether the embryo that eventually grew into me can be considered part of my identity. This is not just a theoretical question in metaphysics. Is the primitive embryo that eventually begot me, the person, part of my identity? If so, destroying such an embryo has moral significance, which destroying another single-celled organism does not have. If the primitive embryo is not part of my identity, then it loses any essential status as part of the continuity of mind that makes me who I am.

PICARD: I agree that you cannot have a memory state of your beginnings as a fertilized egg. But the embryo is part of a continuum that resulted in you, who have a clear identity. A being has moral standing in virtue of its membership in the human species, whether or not it has identity, or whether it has any mental states that recall its existence.

FRANKLIN: Aren't the sperm and egg independently part of the human species? If so, shouldn't they have moral standing?

PICARD: Neither the sperm nor the egg alone can develop into a human being. Neither represents a "human life." I would only consider the diploid cell, the fertilized egg, as having all the necessary components for human life: forty-six chromosomes. Once we recognize that there is moral worth in the primitive human embryo, it follows that the moral worth continues throughout the development of the embryo into the fetus and the child. Whether you recognize it or not, that embryo holds the key to your personal identity.

SPENCER: As I see it, for the primitive embryo that became me to have moral worth, it has to be tied to me by more than a biological continuity through time, where primordial cells beget cells, beget a blastocyst, beget a fetus, begets a person. There has to be a connection to my identity. If the primitive egg split and two people came from it, then there are two human identities, and one egg cannot have both of them. So in the case of the egg before the possibility of splitting, it cannot be a part of my identity.[4]

PICARD: Conjoined twins can have two identities in the same body. So a cell can have two identities until it splits. Your identity is still connected to the primitive egg, whether it split or not.

SPENCER: I do not remember being a single cell or a blastocyst, so my belief that I developed out of a one-celled organism is not based on experiential memory. Since I do not remember being a clump of cells, it is hard for me to think of a clump of source cells as me. The mere fact that the clump and I are spatiotemporally contiguous with each other is not enough to imply that it was me, in other words, that I already existed so many years ago in the form of a clump of cells. The blastocyst, after all, did not have my mind and my consciousness, or the ability to remember anything. In fact, all it had of mine was my DNA, somewhat similar to that of my immediate family.

PICARD: Oh, Jacob! Surely you do not remember your first years after birth either. Science can expand our concept of identity—and I am saying that as a metaphysician. Sequencing the human genome has changed people's concept of self-identity. They seek genetic information on their ancestry. Similarly, as science reveals to us more knowledge of the early embryo, epigenetics, and the human microbiome, it will affect human identity. Would you dismiss your early infancy as not part of your identity?

SPENCER: There is a difference, as far as my identity is concerned, between the early embryo—the blastocyst—and my infancy. The infant has a mind receptive to memories, even if they do not last, unlike the early embryo, which is a bundle of cells with no mind or consciousness, not even a mind in remission like that of a comatose child. It is highly problematic to say that the blastocyst that became you had an identity, since there is no semblance of personhood for which an identity would have meaning.[5]

PICARD: As a neuroscientist, at what point in your embryonic development do you believe your identity began to form?

SPENCER: I would say it was when my "primitive streak," the forerunner to my nervous system, was formed—around fourteen weeks.

PICARD: Okay, I have managed to elicit your first principle about identity in a developing human embryo, and that your conceptus embryo falls outside the range of your personal identity. How do you respond to the claim that if the early embryo that developed into you had been destroyed and your life had never happened, the possibility of your identity would have been eliminated? Because your early embryo made your identity possible, it cannot be dismissed as irrelevant to your identity.

SPENCER: There are many counterfactuals that could have prevented my life from happening—if my mother and father had had sex on another day, or if there had been a miscarriage. I would not have existed under any of these conditions, and therefore there was no identity lost. The embryo is surely a human organism, but it is not a part of human identity. More important, the use of an early human embryo outside of a woman's uterus does not rob anyone of their identity.

FRANKLIN: Jacob, you argue that your early embryo is linked to your physical being but not linked to your personal identity. Isn't that odd?

SPENCER: Not at all. My mother's egg and my father's sperm individually are linked to my physical being, but my personal identity is totally unrelated to them. The material products that contributed to the formation of my being are not part of the memories, mental states, community experiences that form my identity. Because I have moral standing does not imply that all stages or components of my material existence have moral standing.

PICARD: What implications does your theory of identity have for the ethics of embryonic stem cells?

SPENCER: There is an argument that goes something like this. Each person grew from an early embryo, which became fully actualized as a person with human identity. Because the embryo is materially connected to the person, it must share in that person's personal identity, just as a sonogram photograph of the growing fetus is part of the person's personal identity. Therefore, to end the life of the early embryo is to destroy the early identity of a person in development. The embryo gets its moral value from its connection with human identity. This is the argument that I find unpersuasive.

PICARD: Membership in the human species is what gives moral standing. Any human organism, even undeveloped, comatose, or cognitively impaired, can claim membership in the human species even though it cannot speak for itself. It is deserving of moral respect even while not possessing an identity or not being part of the identity of a person. As persons, we can understand phenomenologically why the fertilized egg that gestated into us is part of our identity, just as our identity is expanded by our family and our community.

SPENCER: What makes me the same person I was in my baby picture is that I share some essential property or properties with the baby that renders me a continuous person throughout time. "Individuality" is one such property, which the early blastocyst lacks. Because a human blastocyst and a human being do not share an essential property, they do not share an identity relationship with each other.[6]

PICARD: The essential property in the continuity between you and your blastocyst is your DNA.

SPENCER: But DNA does not establish my individuality. A more reasonable place for pinpointing when an individual human being begins to exist is when the blastocyst no longer has the capacity to cleave into distinct organisms.

PICARD: You have established your ontology of identity based on individuality. I see it as the continuity of being.

SPENCER: I can see your point about how people view the embryo in the uterus through the continuity of DNA. But the fertilized egg in a freezer possesses no human identity because it lacks consciousness or memory, and when outside the uterus, it will never achieve those qualities and therefore cannot be said to possess "lost" or "unrealized" human identity. There is no self-identity in the primitive embryo, and because the research embryo never develops into a human being, there will never be a person whose self-identity includes the embryo.

FRANKLIN: Thank you Jacob and Antoinette. I think you have clarified the reasons behind the response that people have to seeing a very young photo of themselves when they say, "Is that me?" We seem to agree that the very early embryo has no human self-identity, and therefore its use to create stem cells does not destroy an identity in process. The adult person may extrapolate their identity to the early stages of their development, when the material components of their becoming were in a preconscious state.

DIALOGUE 23

EMBRYOS WITHOUT OVARIES

Human eggs and excess embryos have become valuable sources for the development of embryonic stem cells. If restrictions are removed from paying women for their eggs or their discarded embryos contributed to research, and if somatic cell nuclear transfer becomes more prevalent, the demand for oocytes and embryos will be driven up.[1] Imagine for a moment that human embryos or embryolike cells could be produced for use in research as a source of stem cells without using natural human eggs or destroying human embryos. And further imagine that sperm could be produced from the skin cells of a woman. It may sound far-fetched, but biological research in human reproduction continues to introduce new possibilities that were unimaginable a few years ago.

In animal studies scientists have reprogrammed adult skin cells (fibroblasts) by inserting genes for four proteins (transcription factors) that turn them into a pluripotent state so that they behave similar to embryonic stem cells. Some people have called these "embryonic stem cell-like cells." This procedure could obviate the need to begin with human embryos to obtain these pluripotent cells. In a 2007 study by Wernig et al., the authors state: "Notably, the cells [induced reprogrammed stem cells]—derived from mouse fibroblasts—can form viable chimaeras, can contribute to the germ line and can generate live late-term embryos when injected into tetraploid blastocysts."[2] The suggestion is that the pluripotency of the reprogrammed cells can produce any cell in the human body, including gametes.

According to one analysis of the results:

> The stiffest test for pluripotency is that the cells should be capable of forming the germline cells that give rise to future generations. All three groups passed. By injecting the induced pluripotent stem (iPS) cells into early-stage blastocysts, the researchers generated adult mice that included cells derived from iPS cells in all their tissues, including germline cells. Mating these chimeric mice with normal mice produced embryos in which all cells were produced from iPS cells and, for Yamanaka's group, live offspring.[3]

If pluripotent stem cells from adult fibroblasts can produce gametes, what social and ethical implications does that have? Take a skin cell from a woman and turn it into a pluripotent cell and then create an egg cell; do the same for a skin cell from a man and create sperm cells. Combine the two and you have an ersatz embryo; through cell culture techniques, it can yield an endless supply of embryonic stem cells without a natural human egg. None of the experiments has quite done this yet for humans, but the possibility seems close at hand. In a review of studies of in vitro gamete production, Nagano stated: "These studies collectively indicate that the derivation of mammalian germ cells, both female and male, is possible in vitro from pluripotent stem cells."[4] For women who have damaged eggs, say from mitochondrial disease, using iPS might be a way to reprogram their skin cells into normal eggs. According to Testa and Harris, a limiting factor in the development of embryonic stem cells for the study of human disease and therapy is the supply of oocytes. If we created artificial oocytes from induced pluripotent stem cells derived from adult tissues, it would remove a major obstacle to a vast range of beneficial research and therapy.[5] The next dialogue addresses these possibilities and the daunting opportunities for abuse of "unnatural" sperm and eggs.

■ ■ ■

Scene: Francis Davenport is a stem cell scientist who reprogrammed a mouse skin cell into an embryonic stem cell-like cell and from it made fully functioning sperm and egg cells. His goal is to use the same technique for producing human gametes from adult cells. Laura Murphy is a Catholic-educated Ph.D. bioethicist who specializes in reproductive ethics. She recently published a book titled *Embryonic Tyranny: How Science Ignores the Sanctity of Life*. She also writes popular essays on bioethics. Both were invited to appear in the 92nd Street Y lecture series—a popular venue for New Yorkers. The moderator of the event is Dr. Franklin.

FRANKLIN: When I was a child, I was in awe of science. Now, as an adult and a scientist, my awe about what science continues to achieve is orders of magnitude greater than it was. Dr. Davenport, recently your work was internationally acclaimed by the community of stem cell scientists and the international media. You made artificial animal eggs from the skin cells of animals. What prompted you to use induced pluripotent stem cell (iPSC) techniques to make gametes?

DAVENPORT: I should point out that I replicated the extraordinary work of Japanese researchers Hayashi and Saitou at Kyoto University,[6] who created fully functioning spermlike cells from adult mouse skin cells that were reprogrammed to create primordial germ cells. These germ cells are capable of producing eggs or sperm. They used the germlike cells to create eggs and then used the eggs to gestate a healthy mouse offspring.[7]

FRANKLIN: Why would you want to engage in this kind of work?

DAVENPORT: There are two reasons to create gametes from adult cells, one scientific, the other medical.

FRANKLIN: Why don't you start with the scientific purpose first?

DAVENPORT: As you know, American federally funded science has been stifled by restrictions on research using early embryos. There are no signs that these restrictions will be totally removed in the near future. The method of induced pluripotent stem cells does not involve the use or destruction of human embryos, which are produced from the eggs of a woman's ovaries. What we do is reprogram human skin cells to become embryonic-like pluripotent cells.

FRANKLIN: What do you do with those cells?

DAVENPORT: We then activate them to become other cells, including germ cells. In this way we do not have to destroy human embryos, since all of our experimental cells originate from skin cells (fibroblasts).

FRANKLIN: So you can construct eggs and sperm from skin cells?

DAVENPORT: Yes! These are oocytelike and spermlike cells that are produced through the intermediary primordial germ cell-like cells, which have been used to reproduce healthy mice.

FRANKLIN: Why do you use the terms "spermlike" and "egglike"? Are they egg and sperm or not?

DAVENPORT: There may be some differences between natural primordial germ cells and oocytes and those reprogrammed from skin cells. The key is, do they function the same way? The animal experiments so far suggest that they do. Some scientists call the germ cells from reprogrammed somatic cells "primordial germ cell-like cells," or PGCLCs.[8] They are trying to keep the distinction between the artificially produced gametes and the natural ones.

MURPHY: How do you think society will view these experiments that are used to create ersatz sperm, eggs, and embryos?

DAVENPORT: We should not and will not call the parthenogenically activated PGCLCs "embryos." They are artificially created reservoirs for embryonic-like stem cells. If these cells can be used to study disease or for therapeutic purposes, people will embrace them very quickly.

FRANKLIN: What about the medical purposes of your research?

DAVENPORT: There are an increasing number of males with abnormal sperm motility or density. The most prevalent male fertility problem is nonobstructive azoospermia or NOA, caused by testis failure or impaired spermatogenesis. Induced pluripotent stem cell-derived spermatozoa can be a source of male gametes for patients with NOA.[9]

FRANKLIN: What about infertile women? Could this method be used for them as well?

DAVENPORT: Infertility affects over two million women of reproductive age in the United States. If a woman is infertile, we could transplant the nucleus of her skin cell into an enucleated donor egg and induce the diploid egg to divide into a blastocyst. We would then remove and culture the inner cell mass of the blastocyst and foster its differentiation into primordial germ cells, and from these produce haploid oocytes. These oocytes with the woman's chromosomes would then be fertilized by her husband's sperm. However, as a result of the mouse experiments, we might also be able to produce healthy eggs directly from her skin cells by iPSC technology.[10]

MURPHY: This seems so abnormal. How can you get viable sperm and eggs from an adult skin cell? Couldn't there be complications?

DAVENPORT: That is a good question. Merely producing the sperm does not make it ready to be used for producing a child. There have been both encouraging[11] and discouraging indicators of the reprogramming. Researchers have shown that the epigenetic imprints of the sperm cells from the reprogrammed somatic cell were not the same as that from the blastocyst-derived embryonic stem cell.[12] This could lead to phenotypic abnormalities in the offspring. The somatic cells have an epigenetic memory, and that affects their reprogramming into stem cells. But we are at a very early stage, and these issues can be addressed in the future.[13]

FRANKLIN: What about homosexual couples who want a biological child?

DAVENPORT: Two men could potentially have a child in which both parents contribute their genomes. One male contributes sperm through a natural process and the other male, through the lab-assisted process of somatic cell reprogramming, contributes an egg from his X chromosome. Naturally, a woman would have to carry the embryo to term.[14]

MURPHY: How do you get sperm cells from the female with two X chromosomes?

DAVENPORT: The conventional wisdom is that males produce sperm and females make eggs. And that is true, up to a point. From male somatic cells we can, through the use of cell reprogramming, produce embryonic-like stem cells that can be differentiated into germ cells, including eggs. Likewise, from the female somatic cells we can, following the same reprogramming, make sperm, albeit sperm with an X chromosome. The Kyoto technique is more far-reaching than conventional wisdom would have us believe.[15]

FRANKLIN: Dr. Murphy, what do you make of this striking science?

MURPHY: This technique can be used to produce some bizarre offspring. If what you say is correct, then a single male can offer his skin cells to produce a primordial germ cell that can be induced to make an egg and sperm. If the egg is fertilized by the sperm, then one individual can provide both egg and sperm for a child. This is a strange form of reproduction. Is it a form of cloning?

DAVENPORT: As far as I know, this has never been tried on any species. Even as a hypothetical, it is different from cloning, in which the result is a near replica of the genome of the cloned individual. If you use the skin cells of a male to make primordial germ cells, i.e., eggs and sperm, when the two germ cells are combined as in fertilization, there are exchanges (crossover) during meiosis between the two copies of chromosomes. The iPSC gametes (sperm and egg), when combined, will undergo genetic recombination before the egg is fertilized.

MURPHY: So the chromosome of the male donor will be reconfigured from what appeared in the skin cell to the version in its iPSC-derived fertilized egg?

DAVENPORT: Yes. That is why the final iPSC germ cell-derived embryolike cell would not be a clone of the male donor.

FRANKLIN: This has monumental ethical implications for human reproduction. You will not require a woman's DNA to have a child.

MURPHY: Or the male DNA, if it proves successful in human reproduction. With iPSC the woman's skin cell can theoretically produce an egg and X chromosome sperm.

DAVENPORT: Dolly the sheep was cloned by somatic cell nuclear transfer, SCNT. They started with a fertilized egg with its nucleus removed; in the place of the nuclear DNA, they inserted the DNA from cells of the mammary tissue of the original sheep. The process we are talking about does not involve SCNT, but rather reversing the developmental stages of a somatic cell.

MURPHY: Is there a name for this process?

FRANKLIN: I have heard it called "induced pluripotent gametogenesis" or IPG, adding another acronym to the emerging nomenclature.

MURPHY: If this does work in humans, won't it undermine the natural biology of reproduction with unforeseeable consequences?

DAVENPORT: We already have some strange methods of activating an egg to undergo meiosis without sperm, such as parthenogenesis. This has not succeeded in mammals, at least for development to term. We also have among the current techniques of assisted reproduction in vitro fertilization, surrogacy, and donated sperm mechanically introduced into the fallopian tube. If iPSC-produced gametes work, it will simply represent another method of procreation.

MURPHY: Can't you see the psychosocial and cultural problems this technology will create? Imagine that a child was born from a pluripotent-derived sperm and egg from a single individual. This is a kind of autologous reproduction. The child would grow up in a world where there are fathers and mothers; even in cases where the father is anonymous, the child knows there was a male contribution to his birth. A child born by autologous reproduction would be viewed as a freak of nature. You are saying we can get sperm and eggs from a male with X and Y chromosomes and sperm from a female with two X chromosomes.

DAVENPORT: The mice experiments suggest we can obtain egglike and spermlike gametes that so far have been functional in producing offspring. When there are two women donors or a single woman donor, the sperm derived will be an X sperm. The women will not be able to produce a male child.[16]

MURPHY: Will nature permit it?

DAVENPORT: So long as the egg is not missing any critical information from the reprogramming of the somatic cell and there are no epigenetic switches that foul up the process, there is a good chance nature will permit it.

FRANKLIN: What about autologous male offspring?

DAVENPORT: Suppose the skin cell of a male is reprogrammed by induced pluripotency and the primordial germ cells are produced. Next, suppose they are turned into eggs and sperm. Now you are asking, what happens when they are combined?

We are not quite sure, since that experiment has not been done in animals and certainly not in humans. One international group of scientists claims it would be difficult to get viable eggs from male XY cells and sperm from female XX cells.[17] There may be signaling errors that would prevent this type of autologous reproduction.

FRANKLIN: While it is doubtful that induced pluripotent gametogenesis and self-fertilization will be tried or will even work on humans, it behooves us to remember how improbable some events seemed before they were demonstrated. There were many doubts raised about IVF when it first was proposed. Louise Brown, born on July 25, 1978, was considered a freak of nature by some observers, yet she was quite normal in every respect. And today IVF is almost as commonplace as caesarian births and has contributed to the happiness of countless numbers of infertile individuals.

MURPHY: Nothing you have said about using somatic skin cells to produce pluripotent stem cells for tissue repair raises the same ethical or legal questions as does working with human embryos. But once you create eggs without ovaries and sperm without testes, the stakes have changed. The idea of autologous ex vivo procreation is so deviant from nature that I would characterize it as transhumanism. Also, people considering the use of sperm produced by stem cells will have to be fairly confident that it is not abnormal and unsafe.[18] And eventually, there would have to be human trials—if they were even permitted on ethical grounds.

DAVENPORT: You are right. We'll have to have many studies on the safety of the gametes produced if this ever gets applied to humans. We have a few years to think about the ethical implications. The derivation of human eggs and sperm in vitro from iPSCs, in whole or at least in part, is anticipated within fifteen years.[19]

MURPHY: Imagine a not-so-distant future when iPSC becomes so common that it is offered directly to the consumer. A male who cannot produce sperm sends his skin cells to a private laboratory that uses iPSC methods to make sperm, which the man uses to impregnate his wife. Or a homosexual male hires a surrogate to gestate a fertilized egg created from his reprogrammed somatic cells and his partner's sperm.

FRANKLIN: If this process becomes standardized, then a woman will be able to send in some somatic cells she obtained surreptitiously from a man whose child she desires. Without his consent, the laboratory will produce his sperm—maybe not quite his sperm.

MURPHY: How bizarre is that! We'll need new privacy laws to protect surreptitious DNA snipping.

FRANKLIN: As science progresses, its misapplications are boundless. That is why we need civil society. But we should not attempt to address people's failings by proscribing scientific inquiry. There are many positive uses of obtaining gametes from somatic cells. As for autologous reproduction, it is something quite new. Recently a bioethics panel at Japan's Keio University announced its approval for research to create human reproductive cells—sperm and ova—using iPSCs, to treat infertility and other congenital diseases.[20] Once we can convert skin cells to embryonic stem cells, nothing seems to be beyond human reach in science.[21] There is only ethics to hold it back.

DIALOGUE 24

HOW MY CELLS BECAME DRUGS

A company named Regenerative Sciences established a clinic in Broomfield, Colorado, to treat patients with a range of orthopedic conditions, including fractures that failed to heal and chronic bursitis, with a stem cell therapy it called RegenexxTM. The therapy consisted of mesenchymal stem cells taken from the patient's bone marrow and grown in tissue culture for about two weeks. In the human body the stem cells produced bone and cartilage. Company scientists conducted some preliminary treatments, which they claimed helped to treat joint problems in some patients.

According to the Food and Drug Administration (FDA), treatments involving stem cells are drugs and biological products and must comply with its regulatory requirements, including premarket approval. Regenerative Sciences argued that their procedure represented medical practice and lay outside of FDA's jurisdiction. FDA warned Regenerative Sciences that its use of stem cell treatments constituted unlicensed biologics and violated federal law. The FDA brought suit against Regenerative Sciences in the United States District Court for the District of Columbia (*United States v. Regenerative Sciences LLC*).

The next dialogue is a dramatic representation of the legal arguments before the U.S. District Court, District of Columbia. All characters in the dialogue are fictional and the dialogue has been made up, but the positions expressed are summaries of the actual arguments in the case, based on court documents and news analyses.

■ ■ ■

Scene: Courtroom of the U.S. District Court, District of Columbia. Presiding is Judge Sheila Prescott. The plaintiff's attorney, representing the U.S. FDA, is Dennis Phillips; Robert Sentaro is the attorney representing Regenerative Sciences. In attendance, but not representing either legal team, is Dr. Franklin, who is covering the case for a medical journal. The briefs, including amici briefs, with the summary arguments have been presented. Judge Prescott engages the plaintiff and defense attorneys on legal aspects of the case through a series of pointed questions.

JUDGE PRESCOTT: The case before us, Civil Action No. 10–1327, involves the *United States v. Regenerative Sciences LLC.* Scientists at Regenerative Sciences developed a procedure trademarked as RegenexxTM, which uses stem cells to aid patients with various orthopedic ailments. The U.S. Food and Drug Administration undertook an enforcement action against Regenerative Sciences under the authority of the Federal Food, Drug and Cosmetic Act (FDCA), 21 U.S.C. § 301 *et seq.*, and the Public Health Service Act (PHSA), 42 U.S.C. § 201 *et seq.*, and charges them with "causing articles of drug to become adulterated" and "misbranded." Mr. Phillips, are you ready to present your summary argument?

PHILLIPS: I am, your Honor. I am a staff attorney for the Food and Drug Administration. We are filing this suit against Regenerative Sciences because, as we shall argue, the procedures they use to create and administer RegenexxTM violate section 301 of the Federal Food, Drug and Cosmetic Act and section 201 of the Public Health Service Act. The company has been recalcitrant to accept FDA's authority on this matter.

JUDGE PRESCOTT: Mr. Sentaro, are you ready to present your summary response?

SENTARO: I am, your Honor. May it please the court, I am here representing Regenerative Sciences. It is our view that RegenexxTM is a medical procedure that uses the cells of an individual for therapeutic purposes administered exclusively to that individual. It is not a product that the company markets. As a medical procedure, it is not subject to FDA authority, but under the authority of the State of Colorado.[1]

JUDGE PRESCOTT: Mr. Sentaro, can you explain and describe the procedure to which you refer in your brief?

SENTARO: I can, your Honor. A licensed physician, using fine needles, withdraws a small amount of bone marrow or synovial fluid sample from the back of a patient's hip. Blood samples are also drawn from the patient's arm. The samples are sent to Regenerative Science's laboratory, a few miles from the clinic where the samples were taken. A certain group of cells called the mesochymal adult stem cells (MSCs) are isolated from the bone marrow or synovial fluid, cultured, and grown in greater numbers. In this process, natural growth factors are used from the patient's blood.

JUDGE PRESCOTT: What happens to the mesochymal cells after they are produced in your laboratory?

SENTARO: It takes about two weeks to obtain the number of cells we need. Once we have the appropriate cell population, they are combined with an antibiotic doxycycline used to prevent bacterial contamination. Then we send the cells to a laboratory at the University of Colorado, where they undergo quality assurance testing.

JUDGE PRESCOTT: Explain what happens to the cells after they pass the quality assurance test.

SENTARO: The cells are sent to our medical team and then used to treat the patient from whom the original cells were taken. They are injected into the patient's injured area such as their knee, hip, or rotator cuff. The stem cells begin to repair the damaged tissue in the injured area. The procedure is recommended for the treatment of osteoarthritis, nonhealing bone fractures, chronic bulging lumbar discs, and soft tissue injuries.

JUDGE PRESCOTT: How long does it take before improvement is observed?

SENTARO: For many patients, we see improvements between three and six months; for some, improvement can come from one to three months.

JUDGE PRESCOTT: Mr. Phillips, how did FDA first become involved with Regenerative Sciences?

PHILLIPS: We began to see reports on the use of a version of RegenexxTM on the company website. Based on how the company described the use of its cells, FDA sent a letter to Regenerative Sciences dated July 25, 2008, stating that the cell product used in its trademarked RegenexxTM was a *drug* under the Federal Food, Drug and Cosmetic Act. FDA investigators inspected Regenerative Sciences' facilities between February and

April 2009. The inspectors found that the laboratory did not operate in conformity with good manufacturing practices. The laboratories were inspected again in June 2010, and again violations were noted. That's when FDA filed suit. Regenerative Sciences filed a counterclaim charging that FDA was interfering with their medical practice and did not have jurisdiction to regulate its stem cells.

JUDGE PRESCOTT: Mr. Sentaro, how does Regenerative Sciences respond to FDA's allegations about your laboratory?

SENTARO: Your Honor, Regenerative Sciences' laboratory for developing therapeutic stem cells adheres strictly to the International Cellular Medicine Society's (ICMS) professional guidelines and has been audited by independent third parties with no safety concerns reported.

PHILLIPS: May it please the court, Regenerative Sciences has close ties with ICMS. Its CEO was a founding member of the organization and serves on the board of directors. The audit they refer to by ICMS was not an independent, arm's-length review of its facilities. ICMS is an advocacy organization that opposes FDA oversight of stem cells as drugs. There are significant commercial conflicts of interest between the auditing group and Regenerative Sciences.[2]

JUDGE PRESCOTT: Can you point to any deficiencies in the work of Regenerative Sciences?

PHILLIPS: The methods used by Regenerative Sciences to claim success are weak. The sample sizes are very small. There are no controls, placebo or otherwise.

JUDGE PRESCOTT: Mr. Sentaro, how do you view these charges?

SENTARO: FDA's role is appropriate for drugs produced for and administered to many patients—drugs that are standardized and mass produced. These procedures are not appropriate and relevant to evaluate individualized, autologous stem cells. What we do is the opposite of mass production. We are at the frontier of personalized medicine.

PHILLIPS: Whether a stem cell product is mass produced or developed solely for one patient, the cells can be contaminated and carry infectious agents. FDA contends that the RegenexxTM procedure constitutes the manufacturing, holding for sale, and distribution of an unapproved injectable biological drug product, which it refers to as a "cultured cell product."

JUDGE PRESCOTT: Mr. Sentaro, why did Regenerative Sciences challenge FDA's authority?

SENTARO: Your Honor, we do not believe that RegenexxTM is a drug or biological product subject to FDA regulation. It is merely a process, part of a medical practice subject to the laws of the State of Colorado. Since it is only used within the state, FDA lacks jurisdiction over medical practice.[3]

JUDGE PRESCOTT: Mr. Phillips, what is your response?

PHILLIPS: It is not relevant under the law that each product is individualized to a patient.

JUDGE PRESCOTT: How is the term "drug" defined by the Federal Food, Drug and Cosmetic Act?

PHILLIPS: The act defines a drug as an article intended for use in the diagnosis, cure, mitigation, treatment, or prevention of disease in man or other animals, or articles other than food intended to affect the structure or any function of the body of man or other animals. Whether an article is a drug depends on its intended use.[4]

JUDGE PRESCOTT: How does FDA define a biological product?

PHILLIPS: A biological product is defined as a virus, therapeutic serum, toxin, antitoxin, vaccine, blood, blood component, or analogous product applicable to the prevention, treatment, or cure of disease or condition of human beings.

JUDGE PRESCOTT: What about vitamin supplements? Don't they fit your definition?

PHILLIPS: Currently, vitamin supplements are treated like food. They are concentrated amounts of substances found in food. For example, you can eat bananas or take potassium supplements. The cells that Regenerative Sciences are using are not anything like food products. The defendants describe the RegenexxTM procedure as an alternative to traditional surgery for a variety of orthopedic conditions and injuries.

JUDGE PRESCOTT: Mr. Sentaro, how do you respond?

SENTARO: Your Honor, if someone were to have their blood taken and the plasma separated, held in a freezer, and then readministered to the patient after surgery, I do not believe that their blood would be considered a drug under the Food, Drug and Cosmetic Act. That is essentially what Regenerative Sciences is doing. We are drawing blood and bone marrow, culturing the cells, and then reuniting them with the patient. The patient's own cells have curative potencies. These cells have only been minimally manipulated. Our processing does not alter the cells' core biological characteristics. We compare what we do with the practice of in vitro fertilization by a lab or fertility clinic, which is not regulated by FDA.

JUDGE PRESCOTT: How do you respond, Mr. Phillips?

PHILLIPS: If the defendant wants to place RegenexxTM outside drug regulation and call it a human cellular tissue-based product, they would have a lower level of regulatory oversight. But to be regulated as such, it must be minimally manipulated. But their product does not meet that test.

JUDGE PRESCOTT: How does Regenerative Sciences alter the cells?

SENTARO: After being extracted from the patient, the mesochymal cells are isolated and cultured. They are exposed to additives and nutrients, and environmental conditions such as temperature and humidity. We believe this is consistent with minimal manipulation. They are still the patient's cells, and the patient's cells are not drugs. Regenerative Sciences believes that stem cells are body parts and not the property of government or large drug companies.[5]

JUDGE PRESCOTT: Regenerative Sciences has an amicus curiae from the organization Regulatory Relief. Mr. Sentaro, does your client support the amicus?

SENTARO: We do, your Honor.

JUDGE PRESCOTT: Please summarize its position.

SENTARO: If FDA is given regulatory authority over autologous stem cell treatments, the cost of such procedures will skyrocket. Because health insurance companies do not currently cover the procedures, the costs will become out of bounds for all but the very rich. Patients with chronic diseases who would benefit from the procedure believe that FDA is standing in the way of their getting stem cell treatments performed by their physicians. This is causing people to leave the country to obtain these treatments—a response known as stem cell tourism.

JUDGE PRESCOTT: Are there other social responses to the regulation of your process?

SENTARO: If FDA gets regulatory authority over our procedure, it will slow down investment and progress in this significant medical breakthrough.

Small, innovative companies, like Regenerative Sciences, will no longer be players in an overly regulated arena of medical therapy. Only the major drug and medical device companies, known as Big Pharma, will be able to participate. However, they are much more cautious and less prone to risk taking. That means that medical advances will be held back.[6]

And one final point. The doctor-patient relationship has significant value in our society. If FDA is given authority to regulate autologous stem cell transfer, it will interfere with that relationship by restricting a doctor from using a person's own stem cells to treat disease. A person's right to go to a doctor and have their stem cells withdrawn, cultured, and returned to them is not the province of the federal government. Patient rights rest on this decision no less than the rights of small, innovative entrepreneurial companies to survive.

JUDGE PRESCOTT: Your treatment has not been approved as a standard of care, has it?

SENTARO: That is correct. Physicians are always ahead of the curve in discovering effective treatments. The law permits physicians to use their best informed judgment in providing care.

JUDGE PRESCOTT: Even if FDA has regulatory authority over stem cells, that does not prevent physicians' use of the procedure.

SENTARO: FDA's involvement will cast a shadow over the technique; physicians will have to pay higher insurance premiums. FDA is moving toward the regulation of medical practice, which has traditionally been left to the state medical authorities and to local medical boards.

JUDGE PRESCOTT: Thank you, gentlemen. These proceedings have now come to a close.

ACTUAL U.S. DISTRICT COURT DECISION, NOVEMBER 22, 2011[7]

The Defendants state their intentions to use the RegenexxTM procedure for mitigation and treatment of disease and injury. Their statements of intended use fully satisfy the statutory definition for a drug. Also, RegenexxTM, based on mesochymal stem cells derived from the

> patient's bone marrow, fully satisfies the definition of a "biological product" because it is blood, a blood component or derivative, or analogous product applicable to the prevention, treatment or cure of a human disease or condition.
>
> Moreover, I do not accept the claim that RegenexxTM, involving the expansion of mesochymal cells in culture, constitutes minimal manipulation.
>
> Regenerative Sciences is enjoined by this Court from producing and distributing their product RegenexxTM to patients without approval from the Food and Drug Administration by its authority under the Federal Food, Drug and Cosmetic Act.

Regenerative Sciences appealed the District Court decision to the U.S. Court of Appeals. The case was argued on October 21, 2013 and decided on February 4, 2014.

ACTUAL U.S. APPEALS COURT DECISION, FEBRUARY 4, 2014[8]

> We affirm the district court's orders granting summary judgment to the government, dismissing appellants' counterclaims, and permanently enjoining appellants from committing future violations of the FDCA's manufacturing and labeling provisions.

DIALOGUE 25

A CLINICAL TRIAL FOR PARALYSIS TREATMENT

Spinal cord injury (SCI) has been one of the major medical targets of those seeking to advance stem cell research. The late Christopher Reeve and his wife, Dana Reeve, established a foundation with the mission of using stem cells to cure diseases like SCI. According to the Christopher and Dana Reeve Foundation website: "There are at least three basic opportunities presented by embryonic stem cell research. First, it could lead to the development of innovative replacement or transplantation therapies for diseases and disorders such as spinal cord injuries, diabetes, heart disease, and Parkinson's disease."[1] In an interview with June Fox, who worked with Reeve on his book *Still Me,* she was asked:

> Q: In public, Chris always seems so confident that he will be able to walk again, perhaps by his 50th birthday. When he is out of the spotlight, does he seem as sure that a cure for spinal cord injury is near?
>
> June: Absolutely.[2]

Christopher Reeve's paralysis and stem cells became an issue in the 2004 presidential campaign. Vice presidential candidate John Edwards predicted that under a John Kerry presidency, Christopher Reeve would "get out of that wheelchair and walk again."[3] Since then, animal research has continued to make important headway on repairing spinal injury. In 2014, scientists at the University of California, Irvine, transplanted human embryonic stem cell-derived neural precursor cells into mice with induced paralysis and demyelination, the destruction of myelin, an insulating and protective fatty protein that sheaths nerve cells or neurons so that they can function effectively. They reported recovery of the mice from paralysis for at least six months.[4]

The number of people in the United States alive in 2013 with spinal cord injury was around 273,000, and 12,000 people sustain SCI each year. Nearly half of all injuries occurred to individuals between the ages of 16 and 30. The average age of injury was about 43 years, and over 80 percent of the injuries happened among males. Since 2010, the most common cause of SCI have been, in descending order, motor vehicle crashes, falls, and acts of violence.[5]

The spinal cord consists of a thick bundle of nerve cells enclosed in the spinal canal that run from the brain to the lower back. The nerve cells in the spinal cord transmit nerve messages and sensations to and from the brain. The body's spinal fluid and vertebrae help protect the spinal cord. When it is injured, the nerves are unable to transmit signals to and from the brain. Injuries can occur by breaks or dislocations of bones in the vertebrae or crushed discs between the bones that push onto the spinal cord. Nerve cells that have been mildly bruised may start to function again after a period of time. Severely damaged nerves will die and not regenerate. The hope was that pluripotent embryonic stem cells would be a source for replacing damaged nerves of the spinal cord. The definitive evidence was to be found in the clinical trial.

A clinical trial is a set of four scientific tests designed to evaluate the safety and efficacy of medical treatments including drugs, diagnostics, medical devices, and therapy protocols. The results can be presented to the Food and Drug Administration as evidence when applying for a license to market the treatment. Investigators supervising a trial enroll volunteers and/or patients into small pilot studies. There are four stages, called phases. The drug-development process will normally proceed through all four phases over many years. If the drug successfully passes through Phases 0, 1, 2, and 3, it will usually be approved by FDA for use in the general population. Each phase has a different purpose and helps scientists answer a different question.

Phase 0 trials are the first in-human trials. Single subtherapeutic doses of the study drug are given to a small number of subjects (10 to 15) to gather preliminary data on the agent's pharmacodynamics (what the drug does to the body) and pharmacokinetics (what the body does to the drug). In Phase 1 trials, researchers test an experimental drug or treatment in a small group of people (20–80) for the first time to evaluate its safety, determine a safe dosage range, and identify side effects. In Phase 2 trials, the experimental treatment is given to a larger group of people (100–300) to see if it is effective and to further evaluate its safety. In Phase 3 trials, the treatment is given to large groups of people (1,000–3,000) to confirm its effectiveness, monitor side effects, compare it to commonly used treatments or a placebo, and collect information that will allow it to be used safely. Sometimes there are Phase 4 trials (after the product is licensed), called postmarketing studies. In Phase 4 the drug's side effects are studied over a larger population and include the treatment's risks, benefits, and optimal use.

The Geron Corporation of Menlo Park, California, was incorporated in 1990 and initially was devoted to research on aging, but moved into cancer research based on the discovery of telomerase and then expanded to stem cell research. In January 2009, Geron received approval from the FDA for the first human clinical trial to treat patients with thoracic spinal cord injury with cells derived from human embryonic stem cells (hESCs). In August 2009, before any patients were enrolled, FDA placed the trial on hold because preclinical data showed cyst formation at the injury site of animals treated with the drug.

By July 2010, FDA lifted the hold on Geron's clinical trial. The company began enrolling patients for Phase 1, administration of its unique human embryonic stem cells. Geron representatives gave preliminary results of the Phase I trial using their product, GRNOPC1, in October 2011, at a preconference symposium at the joint annual meeting of the American Congress of Rehabilitation Medicine and the American Society of Neuro-Radiology in Atlanta, Georgia. Geron reported that to date, GRNOPC1 had been well tolerated, with no serious adverse events.

Geron noted that their therapeutic product "contains hESC-derived oligodendrocyte progenitor cells that have demonstrated remyelinating, nerve growth stimulating, and angiogenic properties leading to restoration of function in rodent models of acute spinal cord injury."[6] They reported that rodents had extensive remyelination surrounding the rodent axons.

In November 2011, to the surprise of those anticipating an outcome, Geron announced that it was terminating its clinical trial for the treatment of SCI with stem cells and would continue to monitor participants who had been through the trial for fifteen years. Geron stated that the study was too expensive and that the company wanted to allocate its resources elsewhere.

In an article in *Bioethics Forum*, published by the Hastings Center, a nonprofit bioethics institute, Françoise Baylis wrote: "It is one thing to close a trial to further enrollment for scientific reasons, such as a problem with trial design, or for ethical reasons, such as an unanticipated serious risk of harm to participants. It is quite another matter to close a trial for business reasons, such as to improve profit margins."[7] Paralyzed people had joined the trial with the expectation that they might be helped from the stem cell therapy. Ending the trial prematurely violated their contract and destroyed their hopes. Others called it a "textbook case in decision: you dump the earliest candidate with the least certainty, and the highest degree of risk, given the amount of money required."[8] In a public statement, Geron indicated that it had divested all of its stem cell assets to a company named Asterias Biotherapeutics. "Under the terms of the Agreement, upon closing of the transaction, Geron contributed to Asterias intellectual property and tangible assets related to its discontinued human embryonic stem cell programs, including cell lines and a Phase I clinical trial in patients with acute spinal cord injury; intellectual property related to its autologous cellular immunotherapy program, including a Phase I/II clinical trial of autologous immunotherapy in patients with acute myelogenous leukemia; and non-therapeutic applications of pluripotent stem cells, such as cellular assay products for use in drug development and toxicity screening." It appears that the new company took over responsibility for any contractual arrangements associated with the clinical trial.[9]

In 2013, a search of clinicaltrials.gov, a U.S. government voluntary listing of recorded clinical trials, using the search terms "spinal cord injury" and "stem cells" revealed twenty-five ongoing trials for SCI. Seven were in U.S. institutions, including four in private companies; three were in China; three in India; three in Korea; two in Israel; and two in Egypt. As the human clinical trials progress, animal research also continues.

Scientists in the United States and Europe reported neuronal regeneration and improvement of function and mobility in rats impaired by an acute spinal cord injury after stem cell treatment.[10]

The first human clinical trial of applying embryonic stem cells to spinal cord injuries became a highly anticipated event. A YouTube video of a paralyzed rat made to walk after treatment with stem cells added to the drama and optimism for several hundred thousand SCI patients.

Through the voices of fictitious characters, the discussions in the next dialogue reflect publicly available information on the goals and progress of the clinical trial. Dr. Franklin has a great amount of hope that the results could help her father's paralysis. The meeting of these fictitious persons is part of the dramatization. The characters are not intended to resemble any real person. The statements attributed to Dr. Lincoln about Geron are based on the company's public comments about its activities. Any similarities to real people or events are purely coincidental.

■ ■ ■

Scene: The Geron Corporation of Menlo Park, California, was the first company in the United States to have an investigative new drug (IND) using embryonic stem cell-derived cells, named GRNOPC1, approved for a clinical trial on patients with acute spinal cord injury. Fred Lincoln is a medical doctor working for Geron and overseeing the trial. Dr. Franklin has been a consultant to Geron because of her work on stem cells for repairing her father's spinal cord injury. Valerie Legere is a science reporter interviewing Dr. Lincoln about the risks and benefits of the trial, questioning whether there is enough proven science to warrant a test on humans.

LINCOLN: Valerie, welcome to Geron. I would like to introduce you to Dr. Rebecca Franklin, who is a bioethicist and stem cell researcher. She is our consultant on the ethics and science of the trials.

LEGERE: Dr. Lincoln, how was Geron able to get this trial approved as the first FDA-authorized clinical trial involving stem cells? Clearly, there has been a race among biotech firms to go from the lab into clinical trials since the excitement of embryonic stem cells was first reported.

LINCOLN: Geron is a public company that was founded in 1990. The company funded three academic laboratories, and they were among the first to successfully derive embryonic stem cells. We also have eminent academic collaborators. The 2009 Nobel Prize for Physiology and Medicine was awarded to early Geron collaborators, Elizabeth H. Blackburn and Carol W. Greider, along with Jack W. Szostak, for the discovery of how chromosomes are protected by telomeres and the enzyme telomerase. We got there first by sheer persistence, supportive investors, and brilliant collaborators.

LEGERE: How does the clinical trial work?

LINCOLN: This is a Phase 1 trial for an investigative new drug we call GRNOPC1. Its purpose is to assess the safety of administering our stem cell-derived oligodendrocyte progenitor cells into patients suffering from spinal cord injuries.

LEGERE: How many patients will be enrolled in the trial?

LINCOLN: For this phase, we plan to enroll ten patients between the ages of eighteen and sixty-five who have had a spinal cord injury that has paralyzed them from the waist down.

LEGERE: What form of stem cell therapy will they receive?

LINCOLN: The people we enroll must have incurred their spinal cord injury within fourteen days before the trial. We will inject stem cells into the base of their spine, between the seventh and fourteenth day.

LEGERE: Why do you specify that subjects have to have been injured so recently?

LINCOLN: We learned from animal experiments that if we wait too long, scarring occurs, and that will keep the cells from being as effective. We are starting off with the most likely scenario for success, and we need a proof of concept in human subjects.

LEGERE: I understand that a Bay Area patient who suffered a serious spinal cord injury and is paralyzed from the waist down has entered the trial.

LINCOLN: That is correct. We do not reveal any names to protect privacy. This patient was treated by doctors at Stanford Medical School.

LEGERE: What kind of negative outcomes will you be looking for?

LINCOLN: I will let Dr. Franklin take this question.

FRANKLIN: The primary concern is safety. We are looking for any adverse reactions from the injection—from the needle, or any immune-suppression effects from the injected cells. Also, we want to make sure they do not get skin ulcers.

Many of us are concerned that some of these patients will develop ectopic growths, which would be a disaster. Ectopic growths, also known as teratomas, are encapsulated, benign tumors that may grow from human embryonic stem cells. Some proportion of the cells derived from hESCs injected into the body could stray from their intended development pathway and turn into a teratoma. So far, our cells have not formed teratomas in animal studies.[11]

LEGERE: What precautions have you been taking to prevent this?

FRANKLIN: First, the cells we are using are not totally undifferentiated cells, but rather 90 percent of the way toward becoming a glial cell. This proximity to the progenitor cells reduces the probability that a teratoma will be produced. Second, we have extensive purification steps in our manufacturing process so we produce cells that are well characterized.

LEGERE: What kind of positive outcomes will you be looking for?

LINCOLN: We are looking for any return of function or even sensation. If our treatment ever so slightly improves skin sensation and feeling in the lower extremities, or bladder control, we will consider that a success.

LEGERE: How many stem cells will you inject?

LINCOLN: The protocol calls for injecting two million cells. That was the highest dose we used in the animal studies. Of course, we expect the stem cells to reproduce, essentially increasing the dose.

LEGERE: Where did you obtain the stem cells, and did you have to destroy embryos to acquire them?

FRANKLIN: Women volunteered embryos not needed in their fertility treatment. We harvested the embryonic stem cells from those embryos, and in the process the embryos were destroyed.

LEGERE: Why did the FDA stop your first trial in 2009?

LINCOLN: Because this was the first such trial, it was followed very carefully. After the first animal tests, we planned to expand the use of our stem cell product in patients who had both severe and less severe injuries in both the thoracic and cervical regions of the spine. In one of our additional preclinical studies on mice, we observed a higher frequency of animals developing cysts at the injury site that had been seen in other preclinical studies with our stem cell line. We notified the FDA of these results, and the agency placed our clinical trial on hold in August 2009 until we could investigate the cysts. As a result, we developed a set of new assays and ran additional animal studies, where we obtained encouraging positive data. That prompted us to expand our product to a cervical injury model.

LEGERE: Were these cysts a troubling side effect?

LINCOLN: No, they were nonproliferative, confined to the injury site, smaller than the injury cavity, and not associated with adverse effects on the animals. Moreover, cysts much larger in size typically appear in about half of patients with spinal cord injury.

LEGERE: How much has your company invested in this trial?

LINCOLN: The Investigative New Drug Application we filed with the FDA was the largest the agency had ever received—22,500 pages of data. It cost us $45 million to produce. We have already committed about $170 million over more than 15 years to developing the treatment for spinal cord injury.[12]

LEGERE: And what evidence do you have that this could work? Have you seen success in animal studies?

LINCOLN: We funded animal trials on paralyzed rats. The rats regained their ability to move after they were injected with nerve cells made from our stem cells.

LEGERE: What exactly is the mechanism of your treatment?

LINCOLN: A person who has thoracic spinal cord injury experiences demyelination of the nerve cells. Demyelination is the destructive removal of myelin, an insulating and protective fatty protein that sheaths nerve cells or neurons. More specifically, the myelin is wrapped around the long extensions of neurons called axons. Myelin is produced by special glial cells in the central nervous system called oligodendrocytes. Patches of demyelination are known as lesions.

One of our treatments contains human embryonic stem cell-derived oligodendrocyte cells that have demonstrated remyelinating and nerve growth-stimulating properties. In rats, these stem cells have restored function after acute spinal cord injury.[13]

LEGERE: One of the skeptics expressed a concern that the rodent spinal cord and the primate spinal cord differ significantly, both functionally and physiologically. So extrapolating from rodent to human might not work. How do you respond?

LINCOLN: We are required to do preclinical studies on animals. And it probably would have been preferable to do the experiments on monkeys. But there are ethical considerations.

LEGERE: What are those ethical issues?

FRANKLIN: Geron would have had to produce spinal cord injury in monkeys, and that would generate a negative response from animal rights activists and many veterinarians. The cost of the trials would have been a hundred times more. And even if they had been able to use monkeys, they would still have to test it on humans.[14]

LEGERE: Do you have anything to report thus far?

LINCOLN: Geron issued a statement recently that the first two patients are doing well so far and have suffered no serious side effects.

LEGERE: Thank you both for your time. I would like to return for a follow-up interview when the trial is complete.

LINCOLN: My pleasure. I am sure I will have some interesting things to report.

[*Several months go by, and Legere returns for a follow-up interview, this time just with Fred Lincoln.*]

LEGERE: Thank you for making time to see me again, especially during the media blitz on Geron's trial. I read that the company stopped the trial. Usually when we hear things like this, we assume something went wrong—there may have been an adverse effect. Is that what happened?

LINCOLN: The decision to stop the trial had no connection to any effects. The board of directors at Geron had to make a critical financial decision about resource allocation. The company had taken two paths. Geron was in the process of developing an oncology drug that was entering a Phase 2 trial. The other path was continuing their clinical trial on stem cell therapy for spinal cord injury. The company decided to focus on its experimental cancer therapies, which were further along in development. With scarce resources, the company had to shut down the stem cell clinical trial. It is planning to sell or license its stem cell program to another company.

LEGERE: Didn't Geron get funding, or at least an agreement, to borrow up to $25 million from the California Institute for Regenerative Medicine?

LINCOLN: The company did have such an agreement. It returned the $6.5 million it had already borrowed, with interest. The company's board of directors recognized the financial challenges it would face in carrying through the clinical trials and turning the results into a clinical product. It had already incurred considerable expenses in preclinical studies with nearly 2,000 rodents with spinal cord injuries.[15]

LEGERE: What about the patients already being treated?

LINCOLN: The four patients being treated showed no signs of improvement or signs of adverse effects from the stem cells.[16] The company will continue to monitor them. But their hope for a cure has been a casualty of the stem cell politics that has made private research inordinately expensive.

FRANKLIN: Geron's protocols for these trials may very well prove successful for SCI patients. We should not have to depend on the mercurial choices of venture capitalists to discover the answer. There is a public policy responsibility to seek relief for the severely disabled, no less than to find a cure for diabetes or a vaccine for AIDS.

UPDATE

According to a review of the Geron trial by Chapman and Scala in 2012,[17]

> Unexpectedly, in November 2011 Geron announced it was halting the trial and withdrawing from the stem cell field. In a statement, the company claimed its decision was motivated by capital scarcity and uncertain economic conditions and not the lack of promise of stem cell therapies. Geron's recently appointed Chief Executive Officer, who apparently has a different set of priorities than his predecessor, indicated that the company had decided to focus on its novel cancer therapies, which are further along in development. . . . Perhaps Geron was disappointed by the lack of demonstrated efficacy of its therapeutic agents even though it was a Phase I trial.

In 2013 Asterias Biotherapeutics acquired all of Geron's stem cell assets, including its autologous cellular immunotherapy program. Asterias developed a stem cell treatment it named AST-OPC1. On May 22, 2014, *India Pharma News* reported new results from the "first-in-man clinical trial of a cell therapy derived from embryonic stem cells." The trial was conducted by David Apple of the Shephard Center in Atlanta, Georgia; Richard Fessler of Northwestern University in Evanston, Illinois; and Gary Steinberg and Stephen McKenna of Stanford University in Palo Alto, California. In June 2014 *Bloomberg News* reported the success of the Phase 1 trial.[18]

> Asterias Biotherapeutics has announced encouraging results from its Phase I clinical trial assessing the safety of its product, AST-OPC1, in subjects with spinal cord injury. The study represented the first-in-man trial of a cell therapy derived from human embryonic stem cells (hESCs). . . . AST-OPC1 is a population of cells derived from hESCs that contain oligodendrocyte progenitor cells (OPCs). OPCs and oligodendrocytes provide several important supportive functions for nerve cells in the central nervous system. In the Phase I clinical trial, five patients with neurologically complete, thoracic spinal cord injury, as classified by the American Spinal Association Impairment Scale, were administered a relatively low dose of two million AST-OPC1 cells at the spinal cord injury site 7–14 days post-injury. The subjects received low-level immunosuppression for the next 60 days. The patients have been followed to date for 2–3 years through numerous clinical visits, MRIs, and neurological assessments. Delivery of AST-OPC1 was successful in all five subjects with no serious adverse events associated with the intraoperative administration of the cells. In addition, there were no serious adverse events associated with AST-OPC1 itself, or the immunosuppressive regimen. There was no evidence of expanding masses, expanding cysts, infections, cerebrospinal fluid leaks, increased inflammation, or neural tissue deterioration at the injury site of these subjects. Immune monitoring of subjects through one year post-transplantation showed no evidence of antibody-based or cellular immune responses to AST-OPC1, despite complete withdrawal of all immunosuppression at 60 days post-transplant. In four of the five

subjects, serial MRI scans performed throughout the 2–3 year follow-up period indicate that reduced spinal cord cavitation may have occurred and that AST-OPC1 may have had some positive effects in reducing spinal cord tissue deterioration. This effect was seen in the animal model testing of AST-OPC1. There were no unexpected neurological degenerations or improvements in the five subjects in the trial as evaluated by the International Standards for Neurological Classification of Spinal Cord Injury (ISNCSCI) exam. Patients in the trial will be followed for a total of 15 years.

EPILOGUE

When Rebecca Franklin began her journey to gain an understanding of the science and ethics behind stem cells, she was driven by the personal desire to find a cure for her father's paralysis. What she discovered were many intersecting voices expressing the hopes and dreams of the afflicted, as well as the cautions of those following their moral compass on the use and commercialization of embryos in research and medicine. Stem cells brought regenerative medicine from a backwater research project to the forefront of cell biology. Within a decade and a half, a new field of scientific inquiry premised on the regenerative power of embryonic or embryonic-like stem cells was born. Before 1998, there were many efforts to use cells from embryos and fetuses to carry out regenerative functions. Aborted fetuses donated to research provided ample material to pursue this line of research. Reprogramming by nuclear transfer was first accomplished by Briggs and King in 1952,[1] followed by John Gurdon in 1962. Gurdon transferred the nucleus from intestinal cells of adult frogs into unfertilized eggs and was able to produce a tadpole. He demonstrated that even differentiated cells contain the genetic information that is needed for the development of a full organism.

The first extraction of human embryonic stem cells brought new excitement into an old medical agenda. New journals with "stem cell" in the title were created. A flurry of patents were awarded, and Big Pharma was poised for a blockbuster cure. There were 4,547 patents with the term "stem cell" or "embryonic stem cell" awarded by the U.S. Patent and Trademark Office between June 22, 1987, and February 19, 2010.

Every animal study that showed a therapeutic outcome was plastered across the front pages of the popular media outlets. Scores of books ventured into the looking glass, predicting the future of stem cell medicine. More astoundingly, in July 2013, Amazon.com cited 27,334 books with the key words "stem cell" or "embryonic stem cell." Every field of medicine was cited as a potential winner, including dentistry, oncology, cardiology, neurology, ophthalmology, endocrinology, dermatology, and hematology. There was a lot at stake. Huge investments were made by states, the federal government, and the venture capitalist community. The hopes and dreams of millions of people afflicted by disease and debilitation rose on the expectation that treatments and cures were on the horizon. Potential blockbuster drugs and therapies were worth billions of dollars. And there are high stakes in glory and fame for pathbreaking discoveries in medical treatment.

The progress of medical science is replete with stops and starts and functions in small, incremental steps; many treatments that are successful in animals fail when tried on humans. The international collaboration of medical science is among the summum bonum of human civilization. Medical studies are part of a global commons in which the best work is published for everyone to benefit from and build on. Of course, the translation of science into medicine is tied in with intellectual property rights, and that can affect access to drugs and medical treatments. The pendulum of agreement on what is patentable intellectual property can also swing between the extremes of private appropriation and the common good. An example is the 2013 U.S. Supreme Court decision that recalibrated the balance point on the patenting of breast cancer genes (BRCA 1 and BRCA2). The court ruled that unmodified DNA is not patentable material, even if a clever process had been invented for selecting the DNA from the genome.[2]

Some observers have noted the polarization of viewpoints in the stem cell controversies. The Witherspoon Council on Ethics and the Integrity of Science described it as follows:

> The stem cell debates have shown American politics at its best and its worst, with examples both of principled and democratic discourse and plainly dishonest demagoguery. And stem cell research itself has shown

> us science at its most noble and its most debased, with examples of both brilliant researchers pursuing cures from terrible afflictions, and others committing egregious scientific fraud in the hunt for glory.[3]

While I was researching the ethical and scientific debates on stem cells for *Stem Cell Dialogues*, I was acutely aware of the polarized positions. However, I became more interested in the middle ground of controversy, where honest, nuanced discussion and disagreement take place. The *Dialogues* were created to illustrate how evolving science can reframe the debate and create a realignment of positions.

The stem cell controversies represented in this book exhibit some uniquely American ideas about the role of the state, the right to engage in research, and the cultural divide between science supported by public funds and science supported by private funds. It was certainly not the first time that the public and private sectors were allowed to resolve ethical issues on their own terms. During the recombinant DNA controversy in 1975, the NIH established guidelines for transplanting genes from one organism to another that applied exclusively to federal grant recipients. Scientists in the private sector were ostensibly unregulated. The issues at stake were the potential risks of broadening the range of an infectious agent or introducing animal cancer genes into the human gut bacteria. Congress did not see fit to create a single system of regulation or oversight.

A similar situation occurred with human gene therapy, for which a federal oversight committee reviewed research protocols funded by the NIH. The private sector was under no legal obligation to follow the same procedures. This bifurcated model was repeated with respect to stem cells. George Bush's stem cell policy applied exclusively to federal grantees. Others funded by states or the private sector could use any available embryonic stem cell lines.

I have tried to capture in the *Dialogues* the excitement and optimism within the scientific community about the role stem cells would someday play in treating human disease. Whether it was through embryonic stem cells, induced pluripotent stem cells, or nuclear transfer, the enthusiasm among cell biologists was palpable. For example, in 2009 Amabile and Meissner wrote, "Recent developments provide optimism that safe, viral free human iPS cells could be derived routinely in the near future. . . . The approach of generating patient-specific pluripotent cells will undoubtedly transform regenerative medicine in many ways."[4] Their only caveat is that it may take years before all the obstacles to applying stem cells safely and effectively for therapeutic uses are addressed. One of the leading stem cell scientists, Shinya Yamanaka, wrote in 2012, "I believe that iPSC technology is now ready for many applications including stem cell therapies."[5]

Scientists know the stakes are high. Consider just one area—end-stage liver disease, which can be caused by cancer (heptacellular carcinoma) or cirrhosis (most commonly caused by alcoholism, hepatitis B, and hepatitis C). Other than liver transplants, most treatments are not very effective. There are about 18,000 patients in the United States on the waiting list for a liver transplant and only about 4,000 donated cadaver livers available for transplant per year.[6] If part of the damaged liver is removed and replaced by stem cell-derived liver cells (hepatocytes), the liver can be regenerated and victims of end-stage liver disease will have a chance to survive without transplants.

There are two distinct parts of the stem cell research translational program: regenerative medicine and personalized medicine. The *Dialogues* have emphasized both elements. Regenerative medicine is the replacement of damaged human tissue and cells with stem cell-derived differentiated cells. Personalized medicine involves using the diseased person's own adult cells, whether adult stem cells or somatic cells induced to take on an embryonic stem cell-like state, to carry out the regenerative functions. "The application of patient-derived iPSC cells for cell therapy has the advantage of using genetically identical cells, which can be introduced into the patient without the need for immunosuppression."[7] As explained in the introduction, the ideal of modern regenerative medicine is to use the biological resources of one's own body to repair itself or replace worn-out or diseased parts, such as damaged heart or brain cells.[8] This is where science and technology can surpass the limits of natural human evolution—and, some would argue, even aging. Of course, the scientific and medical impulse to transcend every boundary of illness and aging will eventually confront the societal and global impacts of increasing human longevity.[9] Using stem cells for regenerative medicine is simply the latest scientific breakthrough seeking to reach that goal; in the process, it is breaking new ground in dealing with the moral issues raised by medical science and technology. It is worth recalling the words of evolutionary biologist J.B.S. Haldane in his 1928 Conway Memorial Lecture on science and ethics, that "by complicating life, science creates new opportunities of wrongdoing; by altering our worldview, it may lead us into one form or another of ethical nihilism; it can never do us harm by pointing out to us the consequences of our actions."[10]

NOTES

INTRODUCTION

1. Sheldon Krimsky, *Genetic Alchemy* (Cambridge, MA: MIT Press, 1982).
2. F. Sanger and A. R. Coulson, "A Rapid Method for Determining Sequences in DNA by Primed Synthesis with DNA Polymerase." *Journal of Molecular Biology* 94 (3) (May 1975): 441–48. doi:10.1016/0022-2836(75)90213-2. PMID 1100841.
3. Sheldon Krimsky and Tania Simoncelli, *Genetic Justice: DNA Databanks, Criminal Investigation, and Civil Liberties* (New York: Columbia University Press, 2011).
4. Plato, *Euthyphro, Apology, Crito*, introduction by Robert D. Cumming (New York: Bobbs-Merrill, 1956).
5. http://classics.mit.edu/Plato/meno.html (accessed August 20, 2013).
6. Plato, The Meno (380 B.C.), trans. Benjamin Jowett, http://classics.mit.edu/Plato/meno.html (accessed November 6, 2014).
7. Galileo Galilei, *Dialogues Concerning Two New Sciences*, trans. Henry Crew and Alfonso De Salvio (New York: Dover, 1914), 62.
8. Ernan McMullin, ed., *Galileo: Man of Science* (New York: Basic Books, 1967), 152.
9. Sheldon Krimsky, "The Use and Misuse of Critical Gedankenexperimente," *Zeitschrift fur Algemeine Wissenschaftstheorie* 4 (2) (1972): 323–334.
10. Nicolas Malebranche, *Philosophical Selections*, ed. Steven Nadler (Indianapolis: Hackett, 1992), 214.

HARNESSING STEM CELLS FOR REGENERATIVE MEDICINE

1. A. H. Maehle, "Ambiguous Cells: The Emergence of the Stem Cell Concept in the Nineteenth and Twentieth Centuries," *Notes & Records of the Royal Society* 65 (2011): 360.
2. Miguel Ramalho-Santos and Holger Willenbring, "On the Origin of the Term 'Stem Cell,'" *Cell Stem Cell* 1 (1) (June 7, 2007): 35–38.

3. Alexander A. Q. Maximow, "Der Lymphzyt als gemeinsame Stamzelle der schiedenen Blutelemente in der embryonalen Entwicklung und im postfetalen Leben der Säugetiere." Translation: The lymphocyte is a stem cell common to different blood elements in embryonic development and during the postfetal life of mammals. http://www.ctt-journal.com/1-3-en-Maximow-1909.
4. F. R. Sabin, F. R. Miller, K. C. Smithburn, R. M. Thomas, and R. E. Hummel, "Changes in the Bone Marrow and Blood Cells of Developing Rabbits," *Journal of Experimental Medicine* 64 (97) (1936): 115.
5. Frederick R. Applebaum, "Hematopoietic-Cell Transplantation at 50," *New England Journal of Medicine* 357 (October 11, 2007): 1472–1474.
6. Ann B. Parson, *The Proteus Effect* (Washington, DC: Joseph Henry Press, 2004).
7. Max D. Cooper, "Memoriam. Robert A. Good, May 26, 1922–June 13, 2003," *Journal of Immunology* 171 (2003): 6318–6319.
8. Ole Johan Borge, "Alternative Means to Obtain Pluripotent Stem Cells," in *Stem Cells, Human Embryos, and Ethics: Interdisciplinary Perspectives*, ed. Lars Østner, 31 (New York: Springer, 2008).
9. Lars Østner, ed., *Stem Cells, Human Embryos, and Ethics* (New York: Springer, 2008), 31.

DIALOGUE 1. HOPE

1. Spinal cord injury (SCI), also known as myelopathy, can be caused by damage to nerve roots or myelinated fiber tracts that carry signals to and from the brain. Myelin is an electrically insulating biological material made up of proteins and fatty substances that forms a layer (myelin sheath) around the threadlike fibers (axons) of the nerve cell (neuron). The thoracic spinal cord is situated in the T1–T12 thoracic spinal canal. The thoracic vertebral segments form the chest wall and ribs and are the best protected of all the vertebral segments. It takes enormous forces to fracture them. Traumatic injuries of the upper thoracic spinal cord are relatively rare, only 10 to 15 percent of spinal cord injuries (compared to 40 percent cervical, 35 percent thoracolumbar, and 5 percent lumbosacral). Thoracic spinal cord injuries occur as a result of high-speed motor vehicular accidents, tumors that have compressed the spinal cord, and ischemic injuries (insufficient supply of blood) of the spinal cord. They generally are severe and often result in complete loss of neurological function below the injury site.

2. L. M. Bjorklund, R. Sanchez-Pernaute, S. Chung, et al., "Embryonic Stem Cells Develop Into Functional Dopaminergic Neurons After Transplantation in a Parkinson's Rat Model," *Proceedings of the National Academy of Sciences* 99 (2002): 2344–2349.
3. Tara Parker-Pope, "Paralyzed Rats Walk Again, Raising Hopes for Humans," *New York Times,* September 21, 2009.
4. Douglas A. Kerr, Jeronia Liado, Michael J. Shamblott, et al., "Human Embryonic Germ Cell Derivatives Facilitate Motor Recovery of Rats with Diffuse Motor Neuron Injury." *Journal of Neuroscience* 28 (12) (2003): 5131–5140.

5. *MIT Technology Review* reported in 2012: "Neural stem cells, derived from aborted fetal spinal cord tissue, were implanted onto each side of the spinal cord injury in the rats along with a supportive matrix and molecular growth factors. The human stem cells grew into the site of injury and extended delicate cellular projections called axons into the rats' spinal cord, despite the known growth-inhibiting environment of the injured spinal cord. The rats' own neurons sent axons into the transplanted material and the rats were able to move all joints of their hind legs." http://www.technologyreview.com/view/429222/paralyzed-rats-walk-again-after-stem-cell-transplant/ (accessed February 28, 2014).
6. Benedict Carey, "In Rat Experiment, New Hope for Spine Injuries," *New York Times*, May 31, 2012.
7. H. S. Keirstead et al., "Human Embryonic Stem Cell-Derived Oligodendrocyte Progenitor Cell Transplants Remyelinate and Restore Locomotion After Spinal Cord Injury," *Journal of Neuroscience* 25 (19) (2005): 4694–4705.
8. P. Tabakow, G. Raisman, W. Fortuna, et al., "Functional Regeneration of Supraspinal Connections in a Patient with Transected Spinal Cord Following Transplantation of Bulbar Olifactory Ensheathing Cells with Peripheral Nerve Bridging," *Cell Transplantation* 23 (2014) (epub ahead of print).

DIALOGUE 2. WHY IS THIS CELL DIFFERENT FROM OTHER CELLS?

1. Lasker Foundation, http://www.laskerfoundation.org/awards/2005_b_description.htm (accessed February 27, 2014).
2. Lasker Foundation, http://www.laskerfoundation.org/awards/2005_b_description.htm (accessed February 27, 2014).
3. Jonathan Slack, *Stem Cells: A Very Short Introduction* (Oxford: Oxford University Press, 2012), 28.
4. Bhat Ganapathi, "The Status of Stem Cell Therapy in Type-1 Diabetics," *International Journal of Diabetes in Developing Countries* 30 (3) (July–Sept. 2010): 113–117.
5. "The term epigenome is derived from the Greek word epi which literally means 'above' the genome. The epigenome consists of chemical compounds that modify, or mark, the genome in a way that tells it what to do, where to do it, and when to do it. Different cells have different epigenetic marks. These epigenetic marks, which are not part of the DNA itself, can be passed on from cell to cell as cells divide, and from one generation to the next." www.genome.gov/glossary/index.cfm?id=529 (accessed November 15, 2013).
6. A. D. Lander, K. K. Gokofski, F. Y. N. Wan, et al., "Cell Lineages and the Logic of Proliferative Control," *PlOS Biology* 7 (1) (January 2009).
7. Douglas Melton, "Stem Cell Challenges in Biology and Public Policy," *Bulletin of the American Academy* (Summer 2007): 9.
8. N. Amariglio, A. Hirshberg, B. N. Scheithauer, et al., "Donor-Derived Tumor Following Neural Stem Cell Transplantation in Ataxia Telangiectasia Patient," *PLOS Medicine* 6 (2) (February 17, 2009): 221–230.

9. Pascale G. Hess, "Risk of Tumorigenesis in First-in-Human Trials of Embryonic Stem Cell Neural Derivatives: Ethics in the Face of Long-Term Uncertainty," *Accountability in Research* 16 (2009): 175–195.
10. David Cyranoski, "5 Things to Know Before Jumping on the iPS Bandwagon," *Nature* (News Feature) 452 (March 26, 2008): 406–408.
11. See for example, V. Krizhanovsky and S. W. Lowe, "The Promise and Perils of p53," *Nature* 450 (August 27, 2009): 1085; U. Ben-David and N. Benvenisty, "The Tumorigenicity of Human Embryonic and Induced Pluripotent Stem Cells," *Nature Reviews Cancer* 11 (April 2011): 268–277.
12. M. Politis, K. Wu, C. Loane, et al., "Serotonergic Neurons Mediate Dyskinesia Side Effects in Parkinson's Patients with Neural Transplants," *Science Translational Medicine* 238ra 46 (June 30, 2010). DOI:10.1126/scitranslmed.3000976.
13. Eliza Barclay, "Stem-Cell Experts Raise Concern About Medical Tourism," *The Lancet* 373 (March 14, 2009): 884.

DIALOGUE 3. THE PRESIDENT'S STEM CELLS

1. Ethics Advisory Board, Department of Health, Education and Welfare, "Report and Conclusions: HEW Support of Research Involving Human *in Vitro* Fertilization and Embryo Transfer," May 4, 1979.
2. Ibid.
3. John A. Robertson, "Embryo Stem Cell Research: Ten Years of Controversy," *Journal of Law and Medical Ethics* 38 (2) (Summer 2010): 191–203.
4. The original rider can be found in Section 128 of P.L. 104–99. The wording has remained the same each year it has been attached to legislation. For FY2009, the wording in Division F, Section 509 of the Omnibus Appropriations Act, 2009 (enacted March 11, 2009) prohibited HHS, including NIH, from using FY2009 appropriated funds as follows: "SEC. 509. (a) None of the funds made available in this Act may be used for—(1) the creation of a human embryo or embryos for research purposes; or (2) research in which a human embryo or embryos are destroyed, discarded, or knowingly subjected to risk of injury or death greater than that allowed for research on fetuses in utero under 45 CFR 46.208(a)(2) and Section 498(b) of the Public Health Service Act (42 U.S.C. 289g(b)) (Title 42, Section 289g(b), *United States Code*). (b) For purposes of this section, the term 'human embryo or embryos' includes any organism, not protected as a human subject under 45 CFR 46 (the Human Subject Protection regulations) . . . that is derived by fertilization, parthenogenesis, cloning, or any other means from one or more human gametes (sperm or egg) or human diploid cells (cells that have two sets of chromosomes, such as somatic cells)."
5. Story C. Landis, Director, National Institute of Neurological Disorders and Stroke, Department of Health and Human Services, speaking before a Joint Hearing, Committee

on Health, Education, Labor and Pensions, Subcommittee on Health, Human Services, Education and Related Agencies, Committee on Appropriations, U.S. Senate, "Can Congress Fulfill the Promise of Stem Cell Research," January 19, 2007, 24.

6. Similar views were expressed by Leon Kass, Chair, President's Council on Bioethics. Ken Adelman, "Biotechnology & Stem-Cell Research: Interview with Dr. Leon Kass," Washingtonian.com, November 1, 2005, http://www.washingtonian.com/articles/health/biotecnology-stem-cell-research-interview-with-dr-leon-kass/ (accessed April 23, 2014).

DIALOGUE 4. THE DICKEY-WICKER ENIGMA

1. National Institutes of Health, *Report of the Human Embryo Research Panel Ad Hoc Group of Consultants to the Advisory Committee to the Director, NIH*, September 1994.
2. Ibid., xvii.
3. United States Court of Appeals for the District of Columbia Circuit, *James L. Sherley, Dr. and Theresa Delsher, Dr. v. Kathleen Sebelius, Secretary, Department of Health and Human Services.* Appeal from the U.S. District Court of District of Columbia, No. 1:09-cv-01575. The opinion for the court was filed by Chief Judge David Byron Sentelle, with concurring opinions filed by Circuit Judge Karen LeCraft Henderson and Circuit Judge Janice Rogers Brown.
4. http://www.supremecourt.gov/Search.aspx?FileName=/docketfiles/12-454.htm (accessed November 7, 2014).

DIALOGUE 5. THE MORAL STATUS OF EMBRYOS

1. Undraga Schagdarsurengin, Agnieszka Paradowska, and Klaus Steger, "Analysing the Sperm Epigenome: Roles in Early Embryogenesis and Assisted Reproduction," *Nature Reviews Urology* 9 (November 2012): 609–619.
2. National Catholic Bioethics Center. "This declaration expressly leaves aside the question of the moment when the spiritual soul is infused. There is not a unanimous tradition on this point and authors are as yet in disagreement. For some it dates from the first instant; for others it could not at least precede nidation [implantation in the uterus]. It is not within the competence of science to decide between these views, because the existence of an immortal soul is not a question in its field. It is a philosophical problem from which our moral affirmation remains independent. That being said, the moral teaching of the Church is that the human embryo must be treated *as if* it were already ensouled, even if it might not yet be so. It must be treated *as if* it were a person from the moment of conception, even if there exists the theoretical possibility that it might not yet be so. Why this rather subtle, nuanced position,

instead of simply declaring outright that zygotes are ensouled, and therefore are persons? First, because there has never been a unanimous tradition on this point; and second, because the precise timing of ensoulment/personhood of the human embryo is irrelevant to the question of whether or not we may ever destroy such embryos for research or other purposes." http://www.ncbcenter.org/page.aspx?pid=305 (accessed November 15, 2013).

3. Pontifical Academy for Life, "Declaration on the Production and the Scientific and Therapeutic Use of Human Embryonic Stem Cells," August 25, 2000, http://www.vatican.va/roman_curia/pontifical_academies/acdlife/documents/rc_pa_acdlife_doc_20000824_cellule-staminali_en.html (accessed February 10, 2011).
4. Michael J. Sandel, "Embryo Ethics: The Moral Logic of Stem-Cell Research." *New England Journal of Medicine* 351 (July 15, 2004): 207. "For those who adhere to this view [embryos are human beings], extracting stem cells from a blastocyst is morally equivalent to yanking organs from a baby to save other people's lives."
5. Daniel Eisenberg, "Stem Cell Research in Jewish Law," http://www.jlaw.com/Articles/stemcellres.html (accessed February 10, 2011).
6. Elliot N. Dorff, "Embryonic Stem Cell Research: The Jewish Perspective," United Synagogue of Conservative Judaism, http://www.uscj.org/Embryonic_Stem_Cell_5809.html (accessed February 10, 2011).
7. Jeff McMahan, "Killing Embryos for Stem Cell Research," *Metaphilosophy* 38 (2–3) (April 2007): 179.
8. Ibid.
9. Douglas A. Melton, "Stem Cell Challenges in Biology and Public Policy," *Bulletin of the American Academy* (Summer 2007): 6–12.
10. *Davis v. Davis*, 842 SW 2d 588 (June 1, 1992), Tennessee Supreme Court.

DIALOGUE 6. CREATING GOOD FROM IMMORAL ACTS

1. Stephen G. Post, "The Echo of Nuremberg: Nazi Data and Ethics," *Journal of Medical Ethics* 17 (1991): 43.
2. President's Council on Bioethics, *Monitoring Stem Cell Research,* a Report of the President's Council (Washington, D.C.: U.S. Government Printing Office, January 2004), 33.
3. Howard J. Curzer, "The Ethics of Embryonic Stem Cell Research," *Journal of Medicine and Philosophy* 29 (5) (2004): 536.
4. Ibid.
5. Nicholas Wade, "Clinics Hold More Embryos Than Had Been Thought," *New York Times,* May 9, 2003, http://www.nytimes.com/2003/05/09/us/clinics-hold-more-embryos-than-had-been-thought.html?pagewanted=print (accessed April 29, 2014).
6. Curzer 2004, 545.

DIALOGUE 7. CIRCUMVENTING EMBRYOCIDE

1. J. Yu, N. A. Vodyanik, et al., "Induced Pluripotent Stem Cell Lines Derived from Human Somatic Cells," *Science* 318 (2007): 1917–1920.
2. K. Takahashi et al., "Induction of Pluripotent Stem Cells from Adult Human Fibroblasts by Defined Factors," *Cell* 131 (5) (2007): 861–872.
3. Ting Zhou, Christina Binda, Sarah Dunzinger, et al., "Generation of Human Induced Pluripotent Stem Cells from Urine Samples," *Nature Protocols* 7 (12) (2012): 2080–2089.
4. Gina Kolata, "Scientists Bypass Need for Embryo to Get Stem Cells," *New York Times*, November 21, 2007.
5. P. Hou, Y. Li, X. Zhang, et al., "Pluripotent Stem Cells Induced from Mouse Somatic Cells by Small-Molecule Compounds," *Science* 341 (August 9, 2013): 651–654.
6. H. Zhou, S. Wu, J. Y. Joo, et al., "Generation of Induced Pluripotent Stem Cells Using Recombinant Proteins," *Cell Stem Cell* 4 (5) (2009): 381–384.
7. L. Warren, P. D. Manos, T. Ahfeldt, et al., "Highly Efficient Reprogramming to Pluripotency and Directed Differentiation of Human Cells with Synthetic Modified mRNA," *Cell Stem Cell* 7 (5) (2010): 618–630.
8. H. Obokata, T. Wakayama, Y. Sasai, et al., "Stimulus-Triggered Late Conversion of Somatic Cells Into Pluripotency," *Nature* 505 (2014): 641–647.
9. David Cyranoski, "Acid-Bath Stem-Cell Study Under Investigation," *Nature* (February 17, 2014), http://www.nature.com/news/acid-bath-stem-cell-study-under-investigation-1.14738 (accessed March 24, 2014).
10. David Cyranoski, "Lead Author Agrees to Retract Controversial Stem-Cell Paper," *Nature News Blog*, May 28, 2014, http://blogs.nature.com/news/2014/05/lead-author-agrees-to-retract-controversial-stem-cell-paper.html (accessed November 8, 2014).
11. David Cyranoski, "Cell-Induced Stress," *Nature* 511 (July 10, 2014): 140–143.
12. Nikhil Swaminathan, "Stem Cells—This Time Without Cancer," *Scientific American News* (November 30, 2007), http://www.sciam.com/article.cfm?id=stem-cells-without-cancer. Maryann Mott, "Animal-Human Hybrids Spark Controversy," National Geographic News, January 25, 2005, http://news.nationalgeographic.com/news/pf/62295276.html (accessed November 8, 2014).
13. Insoo Hyu, "Stem Cells from Skin Cells: The Ethical Question," *Hastings Center Report* (January–February 2008): 20–22.
14. T.A.L. Bervini, V. Tosetti, M. Crestan, et al., "Derivation and Characterization of Parthenogenic Human Embryonic Stem Cells," paper delivered at the annual meeting of the European Society of Human Reproduction and Embryology, Prague, June 21, 2006.
15. D. Drury-Stewart, M. Song, O. Mohamad, et al., "Highly Efficient Differentiation of Neural Precursors from Human Embryonic Stem Cells and Benefits of Transplantation After Ischemic Stroke in Mice," *Stem Cell Research & Therapy* 4 (2013): 93–106; see also Brevini, Toseti, Crostan, et al., "Derivation and Characterization of Parthenogenic Human Embryonic Stem Cells."

16. Q. Mai, Y. Yu, L. Wang, et al., "Derivation of Human Embryonic Stem Cell Lines from Parthenogenic Blastocysts," *Cell Research* 17 (2007): 1008–1009.
17. S. J. McSweeney and M. D. Schneider, "Virgin Birth: Engineered Heart Muscles from Parthenogenic Stem Cells," *Journal of Clinical Investigation* 123 (3) (March 1, 2013): 1010–1013.

DIALOGUE 8. MY PERSONALIZED BETA CELLS FOR DIABETES

1. K. G. Alberti, and P.Z. Zimmett, "Definition, Diagnosis and Classification of Diabetes Mellitus and Its Complications Part I: Diagnosis and Classification of Diabetes Mellitus Provisional Report of a WHO Consultation," *Diabetic Medicine* 15 (7) (July 1998): 539–553.
2. http://viacyte.com/clinical/clinical-trials/ (accessed November 8, 2014).
3. M. Tachibana, P. Amado, M. Sparman, et al., "Human Embryonic Stem Cells Derived by Somatic Cell Nuclear Transfer," *Cell* 153 (June 6, 2013): 1228–1238.
4. Alex O'Meara, *Chasing Medical Miracles* (New York: Walker, 2009).
5. Gerald D. Fischbach and Ruth L. Fischbach, "Stem Cells: Science, Policy and Ethics," *Journal of Clinical Investigation* 114 (November 2004): 1364–1370.
6. John E. Wagner, Testimony before the Committee on Health, Education, Labor, and Pensions and the Subcommittee on Labor, Health and Human Services, Education and Related Agencies, Committee on Appropriations, U.S. Senate, 110th Congress, "Examining Stem Cell Research, Focusing on Ongoing Federal Support of Both Embryonic and Non-Embryonic Stem Cell Research and Scientific Progress, Including Recent Findings on Amniotic Fluid Stem Cells," January 19, 2007, 34.
7. Veronika S. Urbana, Judit Kiss, János Kovács, et al., "Mesenchymal Stem Cells Cooperate with Bone Marrow Cells in Therapy of Diabetes," *Stem Cells* 26 (1) (January 2008): 244–243.
8. Robert Klitzman and Mark V. Sauer, "Payment of Egg Donors in Stem Cell Research in the USA," *Reproductive Biomedicine* 18 (5) (May 2009): 603–608.
9. George Q. Daley, "Customized Human Embryonic Stem Cells," *Nature Biotechnology* 23 (July 2004): 826–828.

DIALOGUE 9. REPAIRING BRAIN CELLS IN STROKE VICTIMS

1. http://www.cdc.gov/stroke/facts.htm (accessed November 8, 2014).
2. Vivek Misra, Michael M. Ritchie, Laura L. Stone, et al., "Stem Cell Therapy in Ischemic Stroke," *Neurology* 79 (Suppl. 1) (2012): s207–s212.
3. Preeti Sahota and Sean I. Savitz, "Investigational Therapies for Ischemic Stroke: Neuroprotection and Neurorecovery," *Neurotherapeutics* 8 (2011): 434–451.
4. Ibid., 441.
5. Vivek et al., "Stem Cell Therapy in Ischemic Stroke," s209.

6. http://www.renueron.com (accessed March 27, 2014).
7. P. Srivastava, "Restorative Theory in Stroke Using Stem Cells," *Neurology India* 57 (2009): 381–386.
8. Ibid.
9. Nuffield Council on Bioethics, "Intervening in the Brain: Current Understanding and Practices," in *Novel Neurotechnologies* (Nuffield Report, June 2013), 39.
10. Helen K. Smith and Felicity N.E. Gavins, "The Potential of Stem Cell Therapy for Stroke: Is PISCES the Sign?" *The FASEB Journal* 26 (June 2012): 2239–2252.

DIALOGUE 10. REVERSING MACULAR DEGENERATION

1. Natalie D. Bull and Keith R. Martin, "Concise Review: Toward Stem Cell-Based Therapies for Retinal Neurodegenerative Disease," *Stem Cells* 29 (2011): 1170–1175.
2. J. Sima, S. X. Zhang, C. Shao, et al., "The Effect of Angiostatin on Vascular Leakage and VEGF Expression in Rat Retina," *FEBS Letters* 564 (2004): 19–23.
3. Anne B. Parson, *Proteus Effect* (Washington, D.C.: Joseph Henry Press, 2004).
4. V. Tropepe, B. L. Coldes, B. J. Chiasson, et al., "Retinal Stem Cells in the Adult Mammalian Eye," *Science* 287 (5460) (March 17, 2000): 2032–2036. BBC News, "Stem Cells Could Restore Vision," http://news.bbc.co.uk/2/hi/health/3950827.stm (accessed August 13, 2013).
5. D. A. Lambda, A. McUsic, R. K. Hirata, et al., "Generation, Purification and Transplantation of Photoreceptors Derived from Human Induced Pluripotent Stem Cells," *PLOS One* 5 (1) (January 2010): 1–9.
6. S. D. Schwartz, J-P. Hubschman, G. Heilwell, et al., "Embryonic Stem Cell Trials for Macular Degeneration: A Preliminary Report," *The Lancet* online, January 23, 2012, DOI: 10.1016/50140–6736(12)60028–2.
7. Raymond D. Lund, Shaomei Wang, Irina Klimanskaya, et al., "Human Embryonic Stem Cell–Derived Cells Rescue Visual Function in Dystrophic RCS Rats," *Cloning and Stem Cells* 8 (3) (September 29, 2006): 189–199, DOI:10.1089/clo.2006.8.189.
8. V. Marchetti, T. U. Krohne, D. F. Friedlander and M. Friedlander, "Stemming Vision Loss with Stem Cells," *Journal of Clinical Investigation* 120 (9) (September 2010): 3012–3021.
9. Interview with Masayo Takahashi, Knoepfler Lab Stem Cell Blog, http://www.ipscell/2014/01/stem-cell-pioneer-masayo-takahashi-interview-on-ips-cells-clinical-studies-more/ (accessed March 28, 2014).
10. Ferris Jabr, "Sight Seen: Gene Therapy Restores Vision in Both Eyes," *Scientific American* (February 8, 2012), http://www.scientificamerican.com/article/gene-therapy-blindness/?print=true (accessed November 9, 2014).
11. Rajeshwari D. Koilkonda and John Guy, "Leber's Hereditary Optic Neuropathy-Gene Therapy: From Benchtop to Bedside," *Journal of Ophthalmology* 2011, Article ID 179412, 16 pages, 2011. doi:10.1155/2011/179412.

DIALOGUE 11. MY STEM CELLS, MY CANCER

1. Gina Kolata, "Cancers Share Gene Patterns, Studies Affirm," *New York Times*, May 2, 2013, A1.
2. Tara Parker-Pope, "Scientists Urge Narrower Rules to Define Cancer," *New York Times*, July 30, 2013, A1; A12.
3. Robert A. Weinberg, "Coming Full Circle—from Endless Complexity to Simplicity and Back Again," *Cell* 157 (March 27, 2014): 267–271.
4. Jan Paul Medema, "Cancer Stem Cells: The Challenges Ahead," *Nature Cell Biology* 15 (4) (April 2013): 338–344.
5. Bonnie J. Eaves, "Here, There, Everywhere?" *Nature* 456 (4) (2008): 581.
6. Ibid.
7. J. A. Martinez-Climent, E. J. Andreu, and F. Prosper, "Somatic Cells and the Origin of Cancer," *Clinical and Translational Oncology* 9 (September 2006): 647–663.
8. E. Quintana, M. Shackleton, M. S. Sabel, et al., "Efficient Tumour Formation by Single Human Melanoma Cells," *Nature* 456 (4) (December 2008): 593–598.
9. Ibid., 593.
10. Peter B. Dirks, "Stem Cells and Brain Tumours," *Nature* 444 (December 7, 2006): 687–688.
11. Z. Yu, T. G. Pestell, M. P. Lisanti, and R. G. Pestell, "Cancer Stem Cells," *International Journal of Biochemistry and Cell Biology* 44 (12) (December 2012): 2144–2151.
12. M. P. Ablett, J. K. Singh, and Robert B. Clarke, "Stem Cells in Breast Tumours: Are They Ready for the Clinic?" *European Journal of Cancer* 48 (2012): 2104–2106.
13. National Institutes of Health, Stem Cell Information. "Perhaps the most important potential application of human stem cells is the generation of cells and tissues that could be used for cell-based therapies. Today, donated organs and tissues are often used to replace ailing or destroyed tissue, but the need for transplantable tissues and organs far outweighs the available supply. Stem cells, directed to differentiate into specific cell types, offer the possibility of a renewable source of replacement cells and tissues to treat diseases including Alzheimer's diseases, spinal cord injury, stroke, burns, heart disease, diabetes, osteoarthritis, and rheumatoid arthritis." http://stemcells.nih.gov/info/basics/pages/basics6.aspx (accessed April 25, 2014).
14. Sheldon Krimsky, "Eureka! New Ideas in Cell Biology. Review: *The Society of Cells* by Carlos Sonnenschein and Ana Soto," *BioScience* 49 (9) (September 1999): 747–748.
15. Jane E. Visvader and Geoffrey J. Lindeman, "Cancer Stem Cells: Current Status and Evolving Complexities," *Cell Stem Cell* 10 (June 14, 2012): 717–728.
16. Robert A. Weinberg, "Coming Full Circle—From Endless Complexity to Simplicity and Back Again," *Cell* 157(March 27, 2014): 267–271.

DIALOGUE 12. REPROGRAMMING CELLS

1. K. Takahashi and S. Tamanaka, "Induction of Pluripotent Stem Cells from Mouse Embryonic and Adult Fibroblast Cultures by Defined Factors," *Cell* 126(2006): 663–676.
2. Tongbiao Zhao, "Immunogenicity of Induced Pluripotent Stem Cells," *Nature* 474 (7350) (June 2011): 212–215.
3. H. Zhov, S. Wu, J. Y. Joo, et al., "Generation of Induced Pluripotent Stem Cells Using Recombinant Proteins," *Cell Stem Cell* 4 (5) (May 2009): 381–384.
4. Bradley Alicea and José B. Cibelli, "Comparing SCNT-Derived ESCs and iPSCs," in *Mitochondrial DNA, Mitochondria, Disease and Stem Cells,* ed. Justin C. St. John, 465 (New York: Humana Press, 2013).
5. In a 2007 interview in *The New York Times*, Yamanaka said, "When I saw the embryo, I suddenly realized there was such a small difference between it and my daughters. . . . I thought, we can't keep destroying embryos for our research. There must be another way." Martin Fackler, "Risk Taking Is in His Genes," *New York Times*, December 11, 2007. He was also reported as once saying, "If embryo stem cell research is the only way to help patients, then I think that is what we should do. At the same time . . . as a natural feeling, I do want to avoid the usage of human embryos. . . . Human embryos are not like skin cells. They can be babies if transplanted. That is why we are doing what we are doing." And what he did has opened the way to possibly repair or replace damaged or diseased organs. "Yamanaka's Ethical Concern Led to a Greater Discovery . . . and a Nobel Prize," *Japan Daily Press* http://japandailypress.com/yamanakas-ethical-concern-led-to-a-greater-discovery-and-a-nobel-prize-1115448/.
6. Matthias Staatfeld and Konrad Hochedlinger, "Induced Pluripotency: History, Mechanisms, and Applications," *Genes and Development* 24 (2010): 2239–2263.
7. I. Gutierrez-Aranda, V. Ramos-Meijia, C. Bueno, et al., "Human Induced Pluripotent Stem Cells Develop Teratoma More Efficiently and Faster Than Human Embryonic Stem Cells Regardless of the Site of Injection," *Stem Cells* 28 (2010): 1568–1570.
8. P. S. Knoepfler, "Deconstructing Stem Cell Tumorgenicity: A Roadmap to Safe Regenerative Medicine," *Stem Cells* 27 (2009): 1050–1056.
9. Uri Ben-David and Nissim Benvenisty, "The Tumorigenicity of Human Embryonic and Induced Pluripotent Stem Cells," *Nature Reviews/Cancer* 11 (April 2011): 268–277.
10. S. M. Hussein, N. N. Batada, S. Vuoristo, et al., "Copy Number Variation and Selection During Reprogramming to Pluripotency," *Nature* 471(March 2011): 58–64. "Stem Cell Roadblock Discovered in Toronto; Conversion Process Making More Mutations," *Toronto Star*, March 3, 2011.
11. Elie Dolgin, "Flaw in Induced Stem Cell Model," *Nature* 470 (February 3, 2011): 13.
12. H. Ma, R. Morey, R. C. O'Neil, et al., "Abnormalities in Human Pluripotent Cells Due to Reprogramming Mechanisms," *Nature* 511 (July 10, 2014): 177–197.
13. G. J. Sullivan, Y. Bai, J. Fletcher, and I. Wilmut, "Induced Pluripotent Stem Cells: Epigenetic Memories and Practical Implications," *Molecular Human Reproduction* 16 (2) (2010): 880–885.

14. Elie Dolgin, "Flaw in Induced-Stem-Cell Model," *Nature* 470 (February 3, 2011): 13.
15. Martin F. Pera, "The Dark Side of Induced Pluripotency," *Nature* 471 (March 3, 2011): 46–47.
16. J. Bilic, and J.C.I. Belmonte, "Concise Review: Induced Pluripotent Stem Cells Versus Embryonic Stem Cells: Close Enough or Yet Too Far Apart?" *Stem Cells* 30 (11) (2012): 39–40.

DIALOGUE 13. MY PERSONALIZED DISEASE CELLS

1. L. Eide and C. T. McMurray, "Culture of Adult Mouse Neurons," *Biotechniques* 38 (1) (January 2005): 99–104.
2. I-H. Park, N. Arora, H. Huo, et al., "Disease-Specific Induced Pluripotent Stem Cells" *Cell* 134 (September 5, 2008): 877–886.
3. Ibid.
4. Interview of Michael Jackson by John C. Reed, both of the Sanford-Burham Medical Research Institute. "Disease in a Dish: The Ultimate Personalized Medicine," December 7, 2012, http://beaker.sanfordburnham.org/2012/12/disease-in-a-dish-personalized-medicine/ (accessed November 10, 2014).
5. Stephen S. Hall, "Diseases in a Dish," *Scientific American* 304 (3) (March 2011): 41–45.
6. F. Zanella, R. C. Lyon, and F. Sheikh, "Modeling Heart Disease in a Dish: From Somatic Cells to Disease-Relevant Cardiomyocytes," *Trends in Cardiovascular Medicine* 24 (2014): 32–44.
7. Quoted in Alice Park, *The Stem Cell Hope* (London: Plume, 2012), 159.
8. Alice Fano, *Lethal Laws: Animal Testing, Human Health and Environmental Policy* (London: Zed, 1997).

DIALOGUE 14. TO CLONE OR NOT TO CLONE: THAT IS THE QUESTION

1. http://bsp.med.harvard.edu/?q=node/18 (accessed March 6, 2014).
2. J. B. Gurdon, "The Developmental Capacity of Nuclei Taken from Intestinal Epithelium Cells of Feeding Tadpoles," *Journal of Embryology and Experimental Morphology* 10 (1962): 622–640.
3. R. P. Lanza, J. B. Cibelli, and M. D. West, "Human Therapeutic Cloning," *Nature Medicine* 5 (9) (September 1999): 975–977.
4. William J. Broad, "Saga of a Boy Clone Ruled a Hoax," *Science* 211(February 27, 1981): 902. Also, Michael D. Stein, "Rorvik versus Bromhall," *Nature* 296 (April 1, 1982): 383.
5. Transhumanism is the belief or theory that the human race can evolve beyond its current physical and mental limitations, especially by means of science and technology, including implantable electronics, chemical and drug enhancers, and bioengineered babies.
6. Charlotte Kfoury, "Therapeutic Cloning: Promises and Issues," *Montreal Journal of Medicine* 10 (2) (2007): 118.

7. Alexander Meissner and Rudolf Jaenisch, "Mammalian Nuclear Transfer," *Developmental Dynamics* 235 (9) (September 2006): 2460–2469, at 2466.
8. J. B. Cibelli, R. P. Lanza, M. D. West, and C. Ezzell, "The First Human Cloned Embryo," *Scientific American* 235 (9) (January 2002): 45–51, at 47.
9. Alice Park, *The Stem Cell Hope* (London: Plume, 2012), 69.
10. Martin Fackler, "Risk Taking Is in His Genes," *New York Times*, December 11, 2007.
11. Cited in Stuart Newman, "Debating Therapeutic Cloning," *Medical Crossfire* 4 (12) (December 2002): 48–52.
12. L. Turnpenny, S. Brickwood, C. M. Spalluto, et al., "Derivation of Human Embryonic Germ Cells: An Alternative Source of Pluripotent Stem Cells," *Stem Cells* 21 (5) (2003): 598–609.
13. Newman, "Debating Therapeutic Cloning," 51.
14. Gregory J. Downing, "A Researcher's Guide to Federally Funded Human Embryonic Stem Cell Research in the United States," in *Human Embryonic Stem Cells*, ed. Arlene Chiu and Mahendra S. Rao (New York: Springer, 2003), 27–37.
15. Turnpenny, Brickwood, Spalluto, et al., "Derivation of Human Embryonic Germ Cells."
16. Newman, "Debating Therapeutic Cloning," 51.

DIALOGUE 15. PATENTING HUMAN EMBRYONIC STEM CELLS IS IMMORAL AND ILLEGAL (IN EUROPE)

1. Nuala Moran, "European Court Bans Embryonic Stem Cell Patents," *Nature Biotechnology* 29 (2011): 1057–1059.
2. M. A. Bagley, "Patent First, Ask Questions Later: Morality and Biotechnology in Patent Law," *William and Mary Law Review* 45 (2) (January 2004): 469–547.
3. Rick Weiss, "U.S. Denies Patent for a Too-Human Hybrid," *Washington Post*, February 13, 2005, A3.
4. Gretchen Vogel, "Dismay, Confusion Greet Human Stem Cell Patent Ban," *Science* 334 (October 28, 2011): 441.
5. Ibid.

DIALOGUE 16. MY EMBRYO IS AUCTIONED ON THE INTERNET

1. History of eBay, http://www.cs.brandeis.edu/~magnus/ief248a/eBay/history.html (accessed April 4, 2014).
2. Adam Cohen, *The Perfect Store: Inside eBay* (New York: Little, Brown, 2002), 10.
3. Ibid.
4. *Nature Medicine* 12 (May 2006): 487.
5. R. M. Seiderman, C. M. Stojanowski, and F. J. Rich, "The Identification of a Human Skull Recovered from an eBay Sale," *Journal of Forensic Science* 54 (6) (November 2009): 1247–1253.

6. http://www.cnn.com/2014/01/03/us/indianapolis-stolen-brains-ebay/ (accessed November 11, 2014).
7. "Therefore, it would be highly useful to open an experimental window directly on patients' diseased tissue. SCNT could provide a powerful approach. In analogy to 'therapeutic cloning,' the process would require transplanting a somatic nucleus from a patient into an enucleated oocyte to generate a cloned embryo which would be allowed to grow up to the blastocyst stage, at which point ES cells would be derived. . . . The goal here is to have them as an unlimited source from which to derive differentiated diseased cells and diseased tissue as an unprecedented opportunity to understand molecular pathogenesis." G. Testa and J. Harris, "Ethics and Synthetic Gametes," *Bioethics* 19 (2) (2005): 146–166.
8. Donna L. Dickenson, "Tissue Economies: Biomedicine and Commercialization," *Perspectives in Biology and Medicine* 50 (2) (2007): 308–311.
9. The California Stem Cell Agency website lists its goals for ALS: "based on our proven record with successful SCNT in animal studies, we have reasoned that ESC lines can be established from SCNT embryos which will be genetically identical to ALS patients' own cells using frozen human eggs. Upon completion of this project we will be able to prove the concept of SCNT-ESC research for the development of novel therapies for the treatment of human disease. Furthermore, producing such stem cell lines will provide a novel resource to the biomedical research community to study and understand how genes correlate with the development of disease." http://www.cirm.ca.gov/our-progress/awards/establishment-stem-cell-lines-somatic-cell-nuclear-transfer-embryos-humans (accessed November 11, 2014).
10. Julie Wheldon, "Ethical Row Over World's First 'Made to Order' Embryos," *Daily Mail*, August 4, 2006.

DIALOGUE 17. HERE COMES THE EGG MAN: OOCYTES & EMBRYOS.ORG

1. David B. Resnik, "Regulating the Market for Human Eggs," *Bioethics* 15 (1) (2001): 1–25.
2. Rudhika Rao, "Coercion, Commercialization, and Commodification: The Ethics of Compensation for Egg Donors in Stem Cell Research," *Berkeley Technology Law Journal* 21 (3) (June 2006): 1054–1066.
3. Ibid.
4. National Academy of Sciences, Committee on Guidelines for Human Embryonic Stem Cell Research, *Guidelines for Human Embryonic Stem Cell Research* (Washington, D.C.: NAS, 2005).
5. http://www.reproductivefacts.org/; http://stemcell.ny.gov/
6. Resnik, "Regulating the Market for Human Eggs," 22.
7. Kent Sepkowitz, "Is It Ethical to Sell a Human Embryo?" *Newsweek*, June 5, 2013.
8. Scott Carney, "The Cyprus Scramble: An Investigation Into Human Egg Markets," *Pulitzer Center on Crisis Reporting*, August 12, 2012, http://pulitzercenter.org/blog

/untold-stories/cyprus-scramble-investigation-human-egg-markets (accessed April 29, 2014).

9. I. Glenn Cohen and Eli Y. Adashi, "Made-to-Order Embryos for Sale—A Brave New World," *New England Journal of Medicine* 368 (26) (June 27, 2013): 2517–2519.
10. Tarun Jain and Stacey A. Missmer, "Support for Selling Embryos Among Infertility Patients," *Fertility and Sterility* 90 (3) (September 2008): 564–568.
11. Editorial, "400,000 Embryos and Counting," *New York Times*, May 15, 2003, http://www.nytimes.com/2003/05/15/opinion/400,000-embryos-and-counting.html?pagewanted=print (accessed April 29, 2014).
12. Rick Weiss, "400,000 Human Embryos Frozen in U.S.," *Washington Post*, May 8, 2003.
13. Jacqueline Pfeffer Merrill, "Embryos in Limbo," *The New Atlantis* (Spring 2009): 18–28, http://www.thenewatlantis.com/docLib/20090617_TNA24Merrill.pdf (accessed April 29, 2014).
14. K. Pruksananonda, R. Rungsiwiwut, P. Numchaisrika, et al., "Eighteen-year Cryopreservation Does Not Negatively Affect the Pluripotency of Human Embryos: Evidence from Embryonic Stem Cell Derivation," *Bioresearch Open Access* 1 (4) (August 2012): 166–173.

DIALOGUE 18. HUMAN-ANIMAL CHIMERAS AND HYBRIDS

1. Mark Buchan, *Perfidy and Passion: Reintroducing the* Iliad (Madison: University of Wisconsin Press, 2012), 139.
2. Neng Yu, Margot S. Kruskall, Juan J. Yunis, et al., "Disputed Maternity Leading to Identification of Tetragametic Chimerism," *New England Journal of Medicine* 346 (20) (2002): 1545–1552.
3. Maryann Mott, "Animal-Human Hybrids Spark Controversy," *National Geographic News*, January 25, 2005.
4. I. Hyun, P. Taylor, G. Testa, et al., "Ethical Standards for Human-to-Animal Chimera Experiments in Stem Cell Research," *Cell Stem Cell* 1 (August 2007): 159–163.
5. Robert Streiffer, "At the Edge of Humanity: Human Stem Cells, Chimeras, and Moral Status," *Kennedy Institute of Ethics Journal* 15 (4) (December 2005): 350.
6. Stephen S. Hall, *Merchants of Immortality* (Boston: Houghton Mifflin, 2003), 167.
7. Ibid., 171.
8. Cynthia B. Cohen, *Renewing the Stuff of Life* (New York: Oxford University Press, 2007), 120.
9. H. T. Greely, M. K. Cho, L. F. Hogle, and D. M. Satz, "Thinking About the Human Neuron Mouse," *American Journal of Bioethics* 7 (5) (May 2007): 27–40.
10. Nicholas Wade, "Clinton Asks Study of Bid to Form Part-Human, Part-Cow Cells," *New York Times*, November 15, 1998.
11. F. Baylis, "The HFEA Public Consultation Processes on Hybrids and Chimeras: Informed, Effective and Meaningful?" *Kennedy Institute of Ethics Journal* 19 (1) (2009): 41–62.

12. J. S. Robert, "The Science and Ethics of Making Part-Human Animals in Stem Cell Biology," *The FASEB Journal* 20 (May 2006): 840.
13. Greely et al., "Thinking About the Human Neuron Mouse," 27.
14. As discussed by Stephen S. Hall in *Merchants of Immortality*, "The mere notion of a cloning experiment using a cow egg and a human cell seemed guaranteed to violate every commonsense scruple about the natural propriety of scientific inquiry" (171).
15. Streiffer, "At the Edge of Humanity."

DIALOGUE 19. STEM CELL TOURISM

1. Elisabeth Rosenthal, "In Need of a Hip, but Priced Out of the U.S.," *New York Times*, August 4, 2013. See Centers for Disease Control, http://www.cdc.gov/features/medicaltourism/: "'Medical tourism' refers to traveling to another country for medical care. It's estimated that up to 750,000 US residents travel abroad for care each year" (accessed March 10, 2014).
2. Michael D. Horowitz and Jeffrey A. Rosensweig, "Medical Tourism—Health Care in the Global Economy," *The Physician Executive* (November–December 2007): 24–30.
3. Sheldon Krimsky, "China's Gene Therapy Drug," *GeneWatch* 18 (6) (November–December 2005): 10–13.
4. Martha Lagace, "The Rise of Medical Tourism," *Harvard Business School, Working Knowledge*, December 17, 2007.
5. B. D. Colen, "'Stem Cell Tourism' Growing Trend," *Harvard Science*, November 30, 2012, http://news.harvard.edu/gazette/story/2012/11/the-rise-of-stem-cell-tourism (accessed May 29, 2013).
6. "One of the treatments now underway in clinical trials with stem cell therapy is to take a small sample of skin tissue from the burn victim from an area that is not burned and incubate this tissue (about 2 square inches) in an enzyme for about 20 minutes. The enzyme allows the patient's own stem cells to be harvested from the tissue and put in a liquid suspension. From there the cells, suspended in the liquid, are put into a simple spraying device and sprayed onto the affected areas of the burn patient's body. These stem cells spread themselves evenly across the burn wound area and multiply, regenerating the patient's supple skin so that healing takes place quickly and without scarring." http://www.burninjuryfirm.com/stem-cell-research-and-burn-treatment/ (accessed November 17, 2013).
7. "The FDA recognizes only hematopoietic (pertaining to the formation and development of blood cells) stem cell transplantation, corneal resurfacing with limbal stem cells and skin regeneration with epidermal stem cells 'as generally accepted standards of health care,'" http://lifestyle.inquirer.net/128445/stem-cell-therapy-for-burns-not-aging-says-fda (accessed November 17, 2013). S. Dua and A. Azuara-Blanco, "Limbal Stem Cells of the Corneal Epithelium," *Survey of Opthalmology* 44 (5) (March–April 2000): 415–425.
8. Eliza Barclay, "Stem-Cell Experts Raise Concern About Medical Tourism," *The Lancet* 373 (March 14, 2009): 883–884.

9. USFDA, Consumer Health Information, January 2012, www.fda.gov/consumer (accessed April 10, 2014).
10. Kate Kelland, "Health Experts Warn of 'Stem Cell Tourism' Dangers." Reuters, September 1, 2010.
11. Aaron D. Levine, "Stem-Cell Tourism: Assessing the State of Knowledge," *Scripted* 7 (2) (August 2010): 274–282.
12. On May 11, 2011, it was reported that the XCell Center in Germany was closed: "Europe's largest stem cell clinic, which is at the centre of a scandal over the death of a baby given an injection into the brain, has been shut down." Robert Mendick and Allan Hall, "Europe's Largest Stem Cell Clinic Shut Down After Death of Baby," *The Telegraph* (UK), May 8, 2011. The clinic was later reported to have reopened in Lebanon. Robert Mendick, "Stem Cell Doctor Forced to Close His Clinic After Child's Death Is Back in Business," *The Sunday Telegraph*, April 8, 2012.
13. XCell Center, Stem Cell Therapy, http://www.xcell-center.com/ (accessed February 4, 2011).
14. International Society for Stem Cell Reseacrh (ISSCR), Guidelines for the Conduct of Human Embryonic Stem Cell Research, http://www.isscr.org/home/publications/guide-clintrans, (accessed November 17, 2013).
15. Olle Lindvall and Insoo Hyun, "Medical Innovation Versus Stem Cell Tourism," *Science* 324 (June 26, 2009): 1664.
16. D. Lau, U. Ogbogu, B. Taylor, et al., « Stem Cell Clinics Online: The Direct-to-Consumer Portrayal of Stem Cell Medicine," *Cell Stem Cell* 3 (December 4, 2008): 591–594.
17. Kate Kelland, "Health Experts Warn of 'Stem Cell Tourism' Dangers," Reuters, September 1, 2010, http://www.reuters.com/article/2010/09/01/us-stemcells-tourism-idUSTRE67U4VK20100901 (accessed November 12, 2014).

DIALOGUE 20. SOCIAL MEDIA MEET SCIENCE HYPE

1. Zubin Master and David B. Resnik, "Hype and Public Trust in Science," *Science and Engineering Ethics* 19 (2013): 323.
2. Timothy Caulfield, "Biotechnology and the Popular Press: Hype and the Selling of Science," *Trends in Biotechnology* 22 (7) (July 2004): 338.
3. Ibid., 338.
4. Editorial, *Nature Genetics* 35 (1) (September 2003): 1.
5. Caulfield, "Biotechnology and the Popular Press," 337.
6. Editorial, *Nature Genetics* 35 (1) (September 2003): 1.
7. Caulfield, "Biotechnology and the Popular Press," 338.
8. Master and Resnik, "Hype and Public Trust in Science," 324.
9. Nigel Hawkes, "Transplant Teams Grow Cells from Embryos," *The Times of London*, November 6, 1998.
10. Nicholas Wade, "Scientists Cultivate Cells at Root of Human Life," *New York Times*, November 6, 1998.

11. Søren Holm, "Going to the Roots of the Stem Cell Controversy," *Bioethics* 16 (6) (2002): 493–507, at 502.
12. Ibid.
13. Julian Savulescu, "Harm, Ethics Committees and the Gene Therapy Death," *Journal of Medical Ethics* 27 (2001): 148–150.
14. Sheldon Krimsky, *Science in the Private Interest* (Lanham, MD: Rowman & Littlefield, 2003), 132–134.
15. Christine Crofts and Sheldon Krimsky, "Emergence of a Scientific and Commercial Research and Development Infrastructure for Human Gene Therapy," *Human Gene Therapy* 16 (February 2005): 169–177.
16. See "Stem Cell Fraud: A 60 Minutes Investigation," http://www.youtube.com/watch?v=ovPZkQYee8Y (accessed March 10, 2014).
17. Gina Kolata, "Hope in the Lab," *New York Times*, May 3, 1998.

DIALOGUE 21. FEMINISM AND THE COMMERCIALIZATION OF HUMAN EGGS/EMBRYOS

1. Maneesha Deckha, "Legislating Respect: A Pro-Choice Feminist Analysis of Embryo Research Restrictions in Canada," *McGill Law Journal* 58 (1) (2012): 215.
2. Ibid., 217.
3. Deckha, "Legislating Respect," 221.
4. Debora Spar, "The Egg Trade: Making Sense of the Market for Human Oocytes," *New England Journal of Medicine* 356 (13) (March 29, 2007): 1289–1291.
5. Lori Gruen, "Eggs on the Market," *Journal of Ethics in Biology, Engineering and Medicine* 3 (4) (2012): 227–236.
6. Donna L. Dickenson, "The Commercialization of Human Eggs in Mitochondrial Replacement Research," *The New Bioethics* 19 (1) (2013): 18–29, at 22.
7. http://www.rand.org/content/dam/rand/pubs/research_briefs/2005/RB9038.pdf (accessed March 12, 2014).
8. "If the ability to derive 'synthetic' oocytes from ES cells is confirmed also for human ES cells, this could potentially eliminate one of the main caveats of SCNT, namely the supply of human oocytes to perform nuclear reprogramming." G. Testa and J. Harris, "Ethics and Synthetic Gametes," *Bioethics* 19 (2) (2005): 146–166.

DIALOGUE 22. WAS MY BIRTH EMBRYO ME?

1. John Locke, *An Essay Concerning Human Understanding*, vol. 1 (New York: Dover, 1959), 449.
2. Ibid., 450–451.
3. Eric T. Olson, "Was I Ever a Fetus?" *Philosophy and Phenomenological Research* LVII (1) (March 1997): 95–110.

4. Philip J. Nickel, "Ethical Issues in Human Embryonic Stem Cell Research," in *Fundamentals of the Stem Cell Debate*, ed. K. B. Monroe, R. B. Miller, and J. Tobis, 63 (Berkeley: University of California Press, 2008).
5. Ibid., 64.
6. Bertha Alvarez Manninen, "A Metaphysical and Ethical Defense of Human Embryonic Stem Cell Research," *Ethics in Biology, Engineering and Medicine* 3 (4) (2012): 209–225.

DIALOGUE 23. EMBRYOS WITHOUT OVARIES

1. Ruha Benjamin, *People's Science* (Stanford, CA: Stanford University Press, 2013), 91.
2. M. Wernig, A. Meissner, R. Foreman, et al., "*In vitro* Reprogramming of Fibroblasts Into a Pluripotent ES-Cell-Like State."*Nature* 448 (July 19, 2007): 318–325.
3. Jeremy Cherfas, "Three Teams Reprogram Adult Cells For Pluripotency," Archive, ScienceWatch.com, July/August 2008, Archive.sciencewatch.com/ana/hot/bio/08julaug-bio (accessed November 14, 2014).
4. Makoto C. Nagano, "In vitro Gamete Derivation from Pluripotent Stem Cells: Progress and Perspective," *Biology of Reproduction* 76 (2007): 546–551, at 548.
5. Giuseppe Testa and John Harris, "Ethics and Synthetic Gametes," *Bioethics* 19 (2) (2005): 163.
6. K. Hayashi et al., "Offspring from Oocytes Derived from in Vitro Primordial Germ Cell-Like Cells in Mice," *Science* 338 (6109) (November 16, 2012): 971–975.
7. David Cyranoski, "Egg Engineers," *Nature* 500 (August 22, 2013): 392–394; K. Hayashi and M. Saitou, "Generation of Eggs from Mouse Embryonic Stem Cells and Induced Pluripotent Stem Cells," *Nature Protocols* 8 (8) (August 2013): 1513–1524.
8. K. Hayashi and Saitou, "Generation of Eggs from Mouse Embryonic Stem Cells and Induced Pluripotent Stem Cells," *Nature Protocols* 8 (2013): 1513–1524.
9. L. Yao, X. Yu, N. Hui, and S. Liu, "Application of iPS in Assisted Reproductive Technology: Sperm from Somatic Cells?" *Stem Cell Review and Reports* 7 (3) (September 2011): 714–721, at 719.
10. "If stem cell-derived oocytes could someday be safely used for reproductive purposes, a large and diverse group of infertile women would have the opportunity to bear genetically related offspring." C. R. Nicholas, S. L. Chavez, V. L. Baker, et al., "Instructing an Embryonic Stem Cell-Derived Oocyte Fate: Lessons from Endogenous Oogenesis," *Endocrine Reviews* 30 (3) (2009): 277.
11. N. Geijsen, M. Horoschak, K. Kim, J. Gribnau, K. Eggan, and G. Q. Daley, "Derivation of Embryonic Germ Cells and Male Gametes from Embryonic Stem Cells," *Nature* 8 (2004): 106–107.
12. "Epigenetics, literally 'on' genes, refers to all modifications to genes other than changes in the DNA sequence itself. Epigenetic modifications include addition of molecules, like methyl groups, to the DNA backbone. Adding these groups changes the appearance and structure of DNA, altering how a gene can interact with important interpreting (transcribing) molecules in the cell's nucleus. . . . Because they change how genes can

interact with the cell's transcribing machinery, epigenetic modifications, or 'marks,' generally turn genes on or off, allowing or preventing the gene from being used to make a protein. . . . There are different kinds of epigenetic "marks," chemical additions to the genetic sequence. The addition of methyl groups to the DNA backbone is used on some genes to distinguish the gene copy inherited from the father and that inherited from the mother. In this situation, known as 'imprinting,' the marks both distinguish the gene copies and tell the cell which copy to use to make proteins." Johns Hopkins University, Backgrounder: Epigenetics and Imprinting, http://www.hopkinsmedicine.org/press/2002/november/epigenetics.htm.

13. Yao et al., "Application of iPS in Assisted Reproductive Technology."
14. Testa and Harris, "Ethics and Synthetic Gametes."
15. Cyranoski, "Egg Engineers."
16. Hayashi et al., "Offspring from Oocytes Derived from in Vitro Primordial Germ Cell-Like Cells in Mice."
17. Hinxton Group: An International Consortium on Stem Cells, Ethics & Law, http://www.hinxtongroup.org/index.html (accessed March 15, 2014).
18. Abbey Lippman and Stuart Newman, Letter, *Science* 307 (January 28, 2005): 515.
19. The Hinxton Group: An International Consortium on Stem Cells, Ethics & Law, Consensus Statement: Science, Ethics and Policy Changes of Pluripotent Stem Cell-Derived Gametes, April 11, 2008, www.hinxtongroup.org/Consensus_HG (accessed June 23, 2013).
20. "Keio Approves Plan to Create Sperm, Ova Using iPS Cells," February 10, 2011, Daily Yomiuri Online, http://www.yomiuri.co.jp/dy/features/science/T110209004225.htm. http://www.lifeissues.net/writers/irv/irv_194cellasexuallyreproduce2.html (accessed August 26, 2012). https://ajw.asahi.com/article/behind_news/social_affairs/AJ201210060035 (accessed March 16, 2014).
21. Nagano, "In vitro Gamete Derivation from Pluripotent Stem Cells," 550.

DIALOGUE 24. HOW MY CELLS BECAME DRUGS

1. Memorandum of Law in Support of Defendants' Opposition to Plaintiff's Motion to Dismiss, filed April 25, 2011, Civil Action 1:10-CV-01327-RMC, in the U.S. District Court for the District of Columbia, *United States (plaintiff) v. Regenerative Sciences LLC (defendants).*
2. Barbara von Tigerstrom, "The Food and Drug Administration, Regenerative Sciences, and the Regulation of Autologous Stem Cell Therapies," *Food and Drug Law Journal* 66 (2011): 479–506.
3. Ibid.
4. Memorandum of Law in Support of the Plaintiff's Declarations, filed January 7, 2011, Civil Action 1:10-CV-01327-RMC, in the U.S. District Court for the District of Columbia, *United States (plaintiff) v. Regenerative Sciences LLC (defendants).*

5. Memorandum of Defendant's Opposition to Plaintiff's Motion.
6. Von Tigerstrom, "The Food and Drug Administration, Regenerative Sciences, and the Regulation of Autologous Stem Cell Therapies."
7. Memorandum Opinion, filed November 22, 2011, Civil Action No. 10–1327 (RMC), U.S. District Court for the District of Columbia, U.S. District Judge Rosemary M. Collyer.
8. U.S. Court of Appeals, District of Columbia Circuit, No. 12–5254, *United States v. Regenerative Sciences LLC,* decided February 4, 2014, http://www.cadc.uscourts.gov /internet/opinions.nsf/947528CDDA0B9A5A85257C7500533DF4/$file/12-5254-1478137 .pdf (accessed November 15, 2014).

DIALOGUE 25. A CLINICAL TRIAL FOR PARALYSIS TREATMENT

1. http://www.christopherreeve.org/site/c.ddJFKRNoFiG/b.4435095/ (accessed June 27, 2013).
2. http://www.chrisreevehomepage.com/junefox.html (accessed June 27, 2013).
3. Witherspoon Council on Ethics and the Integrity of Science, "The Stem Cell Debates: Lessons for Science and Politics," *The New Atlantic* 34 (Winter 2012): 28.
4. L. Chen, R. Coleman, R. Leang, et al., "Human Neural Precursor Cells Promote Neurologic Recovery in a Viral Model of Multiple Sclerosis," *Stem Cell Report* 2 (2014): 825–837.
5. www.nscisc.uab.edu/publicdocuments/fact_figures_docs/facts%202013.pdf.
6. Geron press release, June 7, 2011, http://ir.geron.com/phoenix.zhtml?c=67323&p=irol -newsArticle&ID=1635754&highlight= (accessed April 14, 2014).
7. Françoise Baylis, "Geron's Discontinued Stem Cell Trial: What About the Research Participants?" *Bioethics Forum,* December 2, 2011, http://www.cnn.com/2004 /ALLPOLITICS/10/12/edwards.stem.cell/.
8. Simon Frantz, "Embryonic Stem Cell Pioneer Geron Exits Field, Cuts Losses," *Nature Biotechnology* 30 (1) (January 2012): 12–13.
9. Geron press release, October 1, 2013, http://ir.geron.com/phoenix.zhtml?c=67323 &p=irol-newsArticle&ID=1860364&highlight= (accessed August 31, 2014).
10. S. van Gorp, M. Leerink, O. Kakinohana, et al., "Amelioration of Motor/Sensory Dysfunction and Spasticity in a Rat Model for Acute Lumbar Spinal Cord Injury by Human Neural Stem Cell Transplantation," *Stem Cell Research and Therapy* 4 (May 28, 2013): 1–22.
11. Joe Alper, "Geron Gets Green Light for Human Trial of ES Cell-Derived Product," *Nature Biotechnology* 27 (March 2009): 213–214.
12. Vicki Glaser, "Interview with Thomas Okarma, CEO, Geron," *Rejuvenation Research* 12 (4) (2009): 295–300.
13. *Journal of Neuroscience* 25 (2005).

14. A. R. Chapman and C.C. Scala, "Evaluating the First-in-Human Clinical Trial of a Human Embryonic Stem Cell-Based Therapy," *Kennedy Institute of Ethics Journal* 22 (3) (2012): 243–261.
15. Frantz, "Embryonic Stem Cell Pioneer Geron Exits Field, Cuts Losses," 12.
16. Andrew Pollack, "Geron Is Shutting Down Its Stem Cell Clinical Trial," *New York Times*, November 15, 2001.
17. Chapman and Scala, "Evaluating the First-in-Human Clinical Trial of a Human Embryonic Stem Cell-Based Therapy," 250.
18. Bloomberg Business Week, June 11, 2014, Asterias Biotherapeutics, http://investing.businessweek.com/research/stocks/private/snapshot.asp?privcapId=234501591.

EPILOGUE

1. R. Briggs and T. J. King, "Transplantation of Living Nuclei from Blastula Cells into Enucleated Frogs' Eggs," *Proceedings of the National Academy of Sciences* 38 (5) (May 1952): 455–463.
2. *Association for Molecular Pathology et al. v. Myriad Genetics, Inc. et al.*, Supreme Court decision, argued April 15, 2013, decided June 13, 2013; https://supreme.justia.com/cases/federal/us/569/12-398/ (accessed November 15, 2014).
3. Witherspoon Council on Ethics and the Integrity of Science, "The Stem Cell Debates: Lessons for Science and Politics," *The New Atlantis* 34 (Winter 2012): 9–60.
4. Giovanni Amabile and Alexander Meissner, "Induced Pluripotent Stem Cells: Current Progress and Potential for Regenerative Medicine," *Trends in Molecular Medicine* 15 (2) (January 2009): 59–68.
5. Shinya Yamanaka, "Induced Pluripotent Stem Cells: Past, Present and Future," *Cell Stem Cell* 10 (June 14, 2012): 678–684.
6. Z. Zhang, J. Liu, Y. Liu, et al., "Generation, Characterization and Potential Therapeutic Applications of Mature and Functional Hepatocytes from Stem Cells," *Journal of Cellular Physiology* 228 (2013): 298–305.
7. G. Hargus, O. Cooper, M. Deleidi, et al., "Differentiated Parkinson Patient-Derived Induced Pluripotent Stem Cells Grow in the Adult Rodent Brain and Reduce Motor Asymmetry in Parkinson Rats," *Proceedings of the National Academy of Sciences* 107 (36) (September 7, 2010): 15924.
8. Marcus-André Deutsch, Anthony Sturzu, and Sean M. Wu, "At a Crossroad: Cell Therapy for Cardiac Repair," *Circulation Research* 112 (2013): 884–890.
9. Stephen S. Hall, *Merchants of Immortality: Chasing the Dream of Human Life Extension* (New York: Houghton Mifflin, 2003).
10. B. S. Haldane, "Science and Ethics," *Conway Memorial Lecture* (London: Watts & Co., 1928).

GLOSSARY

ADULT (ALSO CALLED SOMATIC) STEM CELLS Relatively rare undifferentiated cells found among differentiated cells in tissues or organs. They can renew themselves and can differentiate to yield some or all of the major specialized cell types of the tissue or organ from which they came.

AMYOTROPHIC LATERAL SCLEROSIS (ALS) Sometimes called Lou Gehrig's disease, ALS is a progressive neurodegenerative disease that affects the nerve cells (neurons) in the brain, which control voluntary muscles.

ANEUPLOIDY A condition in which a cell has extra copies or missing copies of specific chromosomes.

ANGIOGENESIS The physiological process through which new blood vessels form. Tumor angiogenesis is the proliferation of a network of blood vessels that penetrates into cancerous growths, supplying nutrients and oxygen and removing waste products.

ANGIOSTATIN A naturally occurring protein (also developed in the laboratory) that inhibits endothelial cells, which are necessary for blood vessel formation. Also an angiogenesis inhibitor.

ARRHYTHMOGENIC DISEASE An abnormal rate of muscle contractions in the heart.

AUTOLOGOUS STEM CELL TRANSPLANT In a stem cell transplant, healthy stem cells are transplanted into a person. In an autologous transplant, the source of the stem cells is the recipient.

AXON The long, threadlike part of a nerve cell along which electrochemical signals are conducted from the cell body to other cells.

BATTEN'S DISEASE A fatal, inherited disorder of the nervous system that typically begins in childhood and is associated with abnormalities in vision, cognitive and speech impairment, and motor disability.

BETA CELLS One of four major types of cells in the pancreas that release insulin into the bloodstream to help regulate blood sugar levels.

CANCER STEM CELLS (CSC) A subpopulation of cancer cells that share many characteristics with normal stem cells, including the ability for self-renewal and differentiation, but also have the ability to regenerate tumors.

CARDIOMYCYTES Cells that make up cardiac muscles in the hearts of vertebrates.

CATEGORICAL IMPERATIVE (IMMANUEL KANT) A standard of rationality and foundational principle of Kant's ethical theory whereby a moral obligation or command is unconditionally and universally binding.

CHOROIDAL NEOVASCULARIZATION (CNV) A condition in which blood vessels in the choroid layer of the eye, beneath the retina, which supply the eye with oxygen and nutrients, break into the retina and disrupt vision.

CHIMERISM A condition in which a person has cells with two distinct sets of DNA. This can occur when two embryos fuse in early pregnancy.

CLONING The process of creating a cell or organism that is genetically identical to the cell or organism from which it was derived.

COMPLEMENTARY DNA (CDNA) A single-stranded segment of DNA synthesized in a laboratory using messenger RNA (mRNA) as a template and the reverse transcriptase enzyme.

DEONTOLOGY The theory of ethics that states that we are morally obligated to act in accordance with a certain set of principles and rules, regardless of the outcome.

DIALYSIS The artificial process of eliminating waste and unwanted water from the blood, which healthy kidneys do naturally.

DIFFERENTIATION The process through which an unspecialized cell, i.e., an embryonic stem cell, acquires the properties of a specialized cell like a nerve, heart, or brain cell.

DIPLOID GENOME A cell or an organism with two sets of chromosomes.

DIZYGOTIC TWINS The product of fertilization of two separate eggs by two separate sperm, as in fraternal twins.

DOPAMINE One of a class of molecules that act as neurotransmitters and hormones and help control the brain's reward and pleasure centers.

DYSKINESIA A condition of diminished voluntary movements and the presence of involuntary movements.

EMBRYONIC STEM CELLS Pluripotent stem cells derived from the inner cell mass of a blastocyst, an early stage embryo, capable of dividing in culture without differentiating for an extended period of time. Human embryos reach the blastocyst stage four to five days after fertilization, at which time they consist of 50 to 150 cells.

ENDOTHELIAL CELLS A thin layer of cells lining the interior surface of all blood vessels, sometimes referred to as the endothelium.

EPIGENETICS The study of gene expression in organisms caused by chemical markings on the DNA—sometimes referred to as chemical switches—rather than alteration of the genetic code itself.

EPITHELIAL CELLS Cells of a type of body tissue known as epithelium, which helps to enclose and protect organs and internal surfaces that have direct contact with outside elements such as food, air, and even sunlight. Found in the lining of the gastrointestinal tract, which has contact with food.

GASTRULATION In the embryonic development of animals, the process of becoming a gastrula, a hollow cup-shaped structure having three layers of cells, the stage following the blastula.

GERM LAYERS The stage of embryonic development after gastrulation when the inner cell mass is organized into three layers of cells: the ectoderm, the mesoderm, and the endoderm.

GLIAL CELLS Non-neural cells in the brain that perform "housekeeper" functions such as clearing out debris and excess materials but do not conduct electrical impulses.

GLIOBLASTOMA The most common malignant brain tumor in humans.

GLYCOPROTEINS A class of compounds in which a protein is combined with a carbohydrate group, commonly found on the surface of cells.

HELA CELLS An immortal cell line initially cultured from cancer patient Henrietta Lacks.

HEMATOPOIESIS The process of production, multiplication, and specialization of all cellular blood components in the bone marrow derived from hematopoietic (blood-forming) stem cells.

HEPATOCELLULAR CARCINOMA Cancer arising from the liver cells (hepatocytes).

HOMUNCULUS A tiny, fully formed individual that (according to the discredited theory of preformation) was supposed to be present in the sperm cell.

INCIDENTALOMAS Lesions or growths that are benign.

INDUCED PLURIPOTENT STEM CELLS Commonly abbreviated as iPS cells or iPSCs; adult cells that have been genetically reprogrammed by the insertion of genes or proteins to induce an embryonic stem cell-like state.

ISCHEMIC STROKE The most common kind of stroke, caused by an interruption in the flow of blood to the brain (as from a clot blocking a blood vessel).

ISLET CELLS Clusters of cells in the pancreas that produce hormones.

KARYOTYPE The number and visual appearance of the chromosomes in the nuclei of a eukaryotic cell.

LAETRILE A chemically modified form of amygdalin, a naturally occurring substance found mainly in the kernels of apricots, peaches, and almonds.

LAMININ A group of proteins that form the structural scaffolding of tissue.

LEBER'S CONGENITAL AMAUROSIS (LCA) An inherited retinal degenerative disease characterized by severe loss of vision at birth.

MACULAR DEGENERATION The progressive deterioration of a critical region of the retina called the macula, resulting in blurred vision.

MESENCHYMAL STEM CELLS Nonblood adult stem cells from a variety of tissues.

MESOCHYMAL CELLS Cells originating in the mesoderm, one of the three primary germ cell layers, that are capable of developing into connective tissues, blood, and lymphatic and blood vessels.

METHYLATION The modification of a strand of DNA after it is replicated, in which a methyl (CH_3) group is added to any cytosine molecule that stands directly before a guanine molecule in the same chain. Since methylation in particular regions of a gene can cause that gene's suppression, DNA methylation is one of the methods used to regulate the expression of genes.

MONADS Indivisible, impenetrable units viewed as the basic elements of physical reality in the metaphysics of the philosopher Leibniz.

MONOZYGOTE One of two or more individuals derived from a single egg, such as an identical twin.

MULTIPOTENT A specialized cell type that is produced from pluripotent stem cells, which can give rise to other types of cells but is limited in its ability to differentiate. For example, blood-forming (hematopoietic) stem cells are multipotent and can differentiate into cell types that are the components of blood.

NEURONS Also called nerve cells; the core components of the nervous system—which includes the brain and spinal cord of the central nervous system (CNS) and the ganglia of the peripheral nervous system (PNS)—that process and transmit information through electrical and chemical signals.

NONOBSTRUCTIVE AZOOSPERMIA (NOA) A condition resulting from the absence or marked reduction of sperm production by the testes. Obstructive azoospermia is due to a blockage in the sperm duct system.

OLIGODENDROCYTE A class of cells in the central nervous system that surrounds and insulates the long fibers (axons) through which the nerves send electrical messages.

ONCOGENESIS The induction or formation of tumors.

OOCYTE A cell that develops into an egg or ovum.

OVARIECTOMY Surgical removal of one or both ovaries; also oophorectomy.

PARTHENOGENESIS A type of asexual reproduction in which a female gamete or egg cell develops into an individual without fertilization from a male gamete. Artificially induced parthenogenesis with human eggs can be used to isolate pluripotent stem cells from a human embryo.

PARTHENOTE An organism produced from an unfertilized ovum (i.e., through parthenogenesis), which is incapable of developing beyond the early embryonic stages.

PETA People for the Ethical Treatment of Animals; an animal rights organization.

PHENOMENOLOGY A philosophical movement devoted to the study of the structure of human consciousness and self-awareness from the first-person point of view. The method emphasizes the immediacy of experience, which is isolated and set apart from all assumptions of existence or causal influence to discover its essential structure.

PHOTORECEPTOR CELLS A specialized type of neurons found in the retina, capable of converting light into signals that can stimulate biological processes.

PLASMID A small circular DNA strand in the cytoplasm of a bacterium or protozoan that can replicate independently of the chromosomes and that is used in genetic engineering to transport genes across different organisms.

PLURIPOTENT STEM CELLS Cells found in an embryo or fetus that can differentiate into all the cell types, with the exception of extra-embryonic tissue such as the trophoblast and placenta.

POLYMERASE CHAIN REACTION A technique in molecular biology for amplifying DNA sequences in vitro that plays a critical role in forensic DNA analysis.

PREIMPLANTATION GENETIC DIAGNOSIS (PGD) In vitro techniques by which embryos are tested for specific genetic or sex-linked disorders before being transplanted into a woman's uterus.

PRIMITIVE STREAK An elongated band of cells that forms along the axis of a developing fertilized egg early in gastrulation. Considered a forerunner of the neural tube and nervous system.

PROGENITOR CELL Like a stem cell, has a tendency to differentiate into a specific type of cell, but is already more specific than a stem cell; it cannot renew itself but can differentiate into its "target" cell.

QUICKENING The moment in pregnancy when the mother starts to feel or perceive fetal movements in the uterus.

RECOMBINANT DNA MOLECULE TECHNOLOGY DNA molecules of different origin that have been joined together by biochemical techniques to make a single molecule in a new genetic combination.

REGENERATIVE MEDICINE Medical therapies involving the use of stem cells to repair, replace, restore, and/or regenerate damaged or diseased cells, tissues, and organs.

REMYELINATION Providing new myelin sheaths to nerve fibers, thereby restoring electrical pathways of the spinal cord.

REPRODUCTIVE CLONING The process of creating in a laboratory a human or animal that is an exact copy of another, using the DNA of the original human or animal.

RESTRICTION ENZYMES Any of a group of enzymes that catalyze the cleavage of DNA at specific sites to produce discrete fragments.

RESTRICTION FRAGMENT LENGTH POLYMORPHISM (RFLP) Any variation in DNA between individuals revealed by restriction enzymes that cut DNA into fragments of different lengths as a result of such variations. RFLP has been used forensically and in the diagnosis of hereditary disease

RETINAL GANGLION CELLS A type of neuron (nerve cell) located near the inner surface of the retina of the eye. They receive visual information from photoreceptors via two intermediate neuron types: bipolar cells and amacrine cells. Damage to these cells may play a role in glaucoma.

RETINAL PIGMENT EPITHELIUM The pigment cell layer located just outside the retina. They are attached to the choroid, the layer of blood vessels and connective tissue between the sclera (white of the eye) and retina, which nourishes the retinal cells.

SOMATIC CELL NUCLEAR TRANSFER (SCNT) A technique that combines an enucleated (nucleus removed) egg and the nucleus of a somatic cell to make an embryo. SCNT can be used for therapeutic or reproductive purposes.

SOMATIC MUTATION THEORY (SMT) The theory that cancer is caused by a single somatic cell that has accumulated multiple DNA mutations and the proliferation of the mutated cells.

SPERMATOGONIAL Pertaining to any of the cells of the gonads in male organisms that are the progenitors of spermatocytes (male germ cells).

STARGARDT MACULAR DYSTROPHY The most common form of inherited juvenile macular degeneration, also called Stargardt disease.

SUPEROVULATION The production of more than the normal number of ova at one time, as through hormone treatment.

TERATOMA A neoplasm (tumor) made up of different types of tissue, none of which is native to the area in which it occurs.

TETRAGAMETIC CHIMERISM The fertilization of two eggs by two spermatozoa followed by the fusion of the zygotes, resulting in a child with intermingled cells from two distinct genomes.

TETRAPLOID BLASTOCYST A blastocyst (embryo), cell, or nucleus with four sets of chromosomes, twice the diploid number in human cells.

THALASSEMIA A group of inherited disorders characterized by reduced or absent amounts of hemoglobin, the oxygen-carrying protein inside the red blood cells.

THERAPEUTIC CLONING The production of embryonic stem cells for use in replacing or repairing damaged tissues or organs, achieved by transferring a diploid nucleus from a somatic cell into an egg whose nucleus has been removed. The stem cells are harvested from the blastocyst that develops from the egg, which, if implanted into a uterus, could produce a clone of the nucleus donor.

THORACIC SPINAL CORD INJURY An injury in the thoracic vertebrae of the mid-back; usually affects the chest and the legs and results in paraplegia.

TISSUE ORGANIZATION FIELD THEORY (TOFT) The theory that carcinogenesis is primarily a problem of tissue organization: carcinogenic agents destroy the normal tissue architecture, disrupting cell-to-cell signaling and compromising genomic integrity.

TISSUE PLASMINOGEN ACTIVATOR (TPA) An enzyme, naturally occurring in small quantities in the blood, involved in the breakdown of blood clots.

T-LYMPHOCYTES Cells that form in the bone marrow and migrate to the thymic cortex; they play a critical role in the immune system.

TOTIPOTENT CELL A nondifferentiated stem cell capable of giving rise to any cell type or of a complete embryo.

TRANSCRIPTION A process during DNA synthesis in which the DNA helix is unwound and one of the strands is read and copied into a new molecule called messenger RNA (mRNA).

TRANSFECTION Infection of a cell with isolated viral nucleic acid followed by production of the complete virus in the cell; also, the incorporation of exogenous DNA into a cell.

TRANSHUMANISM A belief system that claims the human race can transcend its physical, cognitive, and life expectancy limitations by means of science and technology.

TROPHECTODERM Outermost layer of cells in the mammalian blastocyst formed after differentiation of ectoderm, mesoderm, and endoderm.

TUMORIGENESIS The production or formation of tumors.

TYPE 1 DIABETES A condition in which the pancreas produces little or no insulin.

UNIPOTENT STEM CELL A cell capable of differentiating into only one type of cell or tissue.

UTILITARIANISM The philosophical theory that actions are right or moral if they produce the greatest good, happiness, or pleasure.

XENOTRANSPLANTS Procedures in which live cells, tissues, or organs are transplanted from one species to another, i.e., where animal tissues are introduced into the human body.

ZONA PELLUCIDA A transparent glycoprotein outer membrane surrounding the developed mammalian egg.

ZYGOTE The cell that is formed when an ovum (egg) is fertilized.

INDEX

OTHER BOOKS AUTHORED OR EDITED BY SHELDON KRIMSKY

Genetic Alchemy: The Social History of the Recombinant DNA Controversy (1982)
Environmental Hazards: Communicating Risks as a Social Process (with A. Plough) (1988)
Biotechnics and Society: The Rise of Industrial Genetics (1991)
Social Theory of Risk (ed. with D. Golding) (1992)
Agricultural Biotechnology and the Environment (with R. Wrubel) (1996)
Hormonal Chaos: The Scientific and Social Origins of the Environmental Endocrine Hypothesis (2000)
Science in the Private Interest (2003)
Rights and Liberties in the Biotech Age (ed. with P. Shorett) (2005)
Genetic Justice: DNA Databanks, Criminal Investigations, and Civil Liberties (with T. Simoncelli) (2011)
Race and the Genetic Revolution (ed. with K. Sloan) (2011)
Biotechnology in Our Lives (ed. with J. Gruber) (2013)
Genetic Explanations: Sense and Nonsense (ed. with J. Gruber) (2013)
The GMO Deception (ed. with J. Gruber) (2014)